Balfour Stewart

Questions and Exercises on Stewart's Lessons in Elementary Physics

Balfour Stewart

Questions and Exercises on Stewart's Lessons in Elementary Physics

ISBN/EAN: 9783337012700

Printed in Europe, USA, Canada, Australia, Japan

Cover: Foto ©berggeist007 / pixelio.de

More available books at **www.hansebooks.com**

QUESTIONS AND EXERCISES

ON

STEWART'S LESSONS

IN

ELEMENTARY PHYSICS.

BY

GEORGE A. HILL,

ASSISTANT PROFESSOR OF PHYSICS IN HARVARD UNIVERSITY.

WITH ANSWERS AND OCCASIONAL SOLUTIONS.

BOSTON:
GINN AND HEATH.
1880.

Entered according to Act of Congress, in the year 1874,
BY GINN BROTHERS,
in the Office of the Librarian of Congress, at Washington.

GIFT OF
ENGINEERING LIBRARY

PRESS OF ROCKWELL AND CHURCHILL,
39 Arch St., Boston.

PREFACE.

The following pages have been drawn up with the aim of making Mr. Stewart's excellent work more useful in elementary teaching.

Part I. consists of questions upon the text of Mr. Stewart's book which are intended to be direct and exhaustive. Opinions differ as to the value of such questions. No doubt a thoroughly competent teacher will ask questions in his own way with the best effect; but unfortunately such teachers, at least in scientific subjects, are not numerous. In all cases the questions will be found useful for review and examination purposes.

Parts II. and III., which form the principal part of the work, have been written with two objects in view. First, to stimulate original thought on the part of the student, and to give the teacher the means of testing thoroughly the student's knowledge of principles. Secondly, to make certain needful additions to the felicitous but cursory sketch of Mechanics, Hydrostatics, and Pneumatics, contained in the first two chapters of Mr. Stewart's book.

Molecular Physics is rapidly assuming the character of an exact science; and in proportion as this takes place, the importance of a good knowledge of the general laws of Motion and Force, and of the ability to reason deductively, increases. Nothing can give training in deduction better than the study of Rational Mechanics. Training in the methods of induction, which is so large a part of scientific culture, cannot, in our judgment, be imparted successfully by the study of text-books; the place to receive it is in Physical Laboratories, which are happily becoming more and more common, or by observation and reflection in the vast Laboratory of Nature around us. The chief value, which the text-book study of Physics can be made to have, consists in disciplining the mind in scientific demonstration of the deductive kind.

The Exercises are divided into two classes, as explained on page 69. A few of them are original; the most have been selected from English works. Few of them require much numerical work, and many of them none at all.

In preparing the Solutions, the author has been under obligations to the elementary writings of Professors Thomson, Tait, and Maxwell.

It was found impossible to prepare solutions of the more difficult Exercises in small type in season for the present edition. In working out these Exercises, where aid is found necessary it should be obtained, if possible, from a competent teacher. The student is also strongly advised to consult special works which treat of the subjects covered by the Exercises. By doing this, the student will not merely find the aid which he desires, — he will be acquiring a habit of mind which is characteristic of the cultivated man and of all productive scholarship, the habit of consulting and carefully comparing the views which different minds take of the same subject, and of that originality in thought which comes from an independent use of many authorities. On page 69 will be found a list of elementary works which may be consulted with advantage.

G. A. HILL.

CAMBRIDGE, August 25, 1874.

CONTENTS.

		PAGE
PART I.	QUESTIONS	1–68
II.	EXERCISES AND PROBLEMS	69–114
III.	ANSWERS AND SOLUTIONS	115–168
	Answers	115
	Solutions	127
APPENDIX		169–188
I.	English Weights and Measures	169
II.	United States Weights and Measures	173
III.	Metric Weights and Measures	175
IV.	Mathematical Formulae	179
V.	Physical Tables	185

PART I.

QUESTIONS ON STEWART'S ELEMENTARY PHYSICS.

Introduction.

1. How do we become aware of the existence of objects outside of ourselves?

2. What is the ground of our expectation that the sun will rise to-morrow? In general, when is our expectation that a certain phenomenon will recur well grounded?

3. What are characteristics of the knowledge of physical laws which men acquire from every-day experience?

4. How long since men first set themselves systematically to the task of acquiring a knowledge of the laws of nature?

5. What is the object of Physics?

6. What do we learn from astronomy concerning the magnitude of the Universe?

7. Explain the three-fold division of matter into *substances, molecules,* and *atoms.* Illustrate by a familiar example.

8. What is the analogous three-fold division in astronomy?

9. What resemblance exists between the structure of the Universe and that of a body on the earth's surface, in consequence of which both may be called *porous*?

10. Distinguish between physical and sensible pores. What proves the existence of physical pores? Give instances of bodies having sensible pores.

11. Name the three states of matter. What are their chief characteristics? Give examples of each state.

12. Explain relative motion by means of the motions of the earth and the planets.

13. What have we strong reason to suppose to be the condition of any substance at rest, — the molecules of a block of stone, for example?

14. In view of our present scientific knowledge, what may be asserted of every body in the universe, large or small?

15. Illustrate by examples the meaning of the term *force.*

16. Mention three of the most universal forces of nature. What are their effects? What consequences would ensue if these forces severally were to cease to exist?

17. What must be the effect of a single force acting upon a body? What *may* be the effect of two or more forces acting simultaneously on a body? Illustrate by the force of gravitation.

CHAPTER I.

LAWS OF MOTION.

LESSON I. — *Determination of Units.*

18. What is the unit of time?

19. Mention the two chief advantages of the Metric system of weights and measures.

20. What is the unit of length in the Metric system, and its value approximately in English inches?

21. Enumerate the chief multiples and sub-multiples of the metre, and give their values in terms of the metre.

22. How are units of surface and of capacity derived from those of length? Give examples of each kind.

23. What is the value in square metres of an *are?* of a *centiare?* of a *hectare?*

24. What is a *litre?* Name some of its decimal multiples and sub-multiples.

25. What is the ratio between two successive units of length, as the centimetre and the decimetre? between the two corresponding units of surface? between the two corresponding units of volume?

26. How many square feet are there in 150 square inches? How many square centimetres are there in 150 square millimetres?

27. How many cubic yards are there in 93 cubic feet? How many litres are there in 1789 millilitres?

28. What superiority of the Metric system is established by such questions as **26** and **27**?

29. What is the unit of mass in the Metric system, and how is it connected with the unit of length?

30. Enumerate the chief derivative units of mass, and give their values in terms of the gramme.

31. Illustrate the meaning of *velocity* by the example of a railway train. How would you define the word?

32. Show that the space passed over by a body moving for any time with a uniform velocity is equal to the velocity multiplied by the time.

33. What is a convenient unit of velocity?

34. How may the relative masses of bodies of the same kind be estimated?

35. Why cannot *weight* be adopted as a fundamental method of measuring mass?

36. What is the ultimate test that two different substances have the same mass?

37. What relation between mass and weight has been established which enables us to employ weight as a convenient practical means of estimating mass?

38. Define the unit of force.

39. It is true that it requires a double force to produce either (1) the same velocity in a double mass, or (2) a double velocity in the same mass; show that one of these truths follows immediately from the definition of the unit of force?

LESSON II. — *First Law of Motion.*

40. What does the first law of motion assert?

41. Explain how this law is apparently, but not really, contradicted by every-day experience.

42. What are the two great forces which tend to stop all motion on the surface of the earth? Give illustrations of each.

43. What is the nearest approach to a perpetual motion with which we are acquainted?

44. Explain the following illustrations of the first law of motion:—

1. A man is on horseback, and the horse starts off suddenly. In what direction will the man fall?

2. A man is on horseback, and the horse stops suddenly. In what direction will the man fall?

45. Show how the first law of motion serves to explain some of the common phenomena of rotation.

LESSON III. — *Second Law of Motion.*

46. State the second law of motion.
47. For the sake of clearness, what two cases may be considered separately under this law ?
48. Suppose that a ball is thrown upwards or sideways in a moving railway-carriage ; show that its motion relative to the carriage is different from its motion relative to the ground, and that the motion relative to the ground is represented by the diagonal of a parallelogram, the sides of which represent the motions of the ball and of the carriage respectively.
49. If I leap vertically upwards at the equator, I alight upon the place from which I sprang, although all places on the equator are moving, in consequence of the earth's rotation, at the rate of about one mile in three seconds ; explain this by means of the first and second laws of motion.
50. A balloon at the height of two miles above the earth's surface is totally immersed in, and carried along with, a current of air moving at the rate of 60 miles an hour. A feather is dropped over the edge of the car ; will it be blown away, or will it appear to drop vertically down ?
51. A ship is in rapid motion, and a stone is dropped from the top of the mast ; where will it fall ?
52. Examine the case in which a force produces motion in the same direction as an already existing motion, as when a ball is thrown directly forwards in a moving railway-carriage.
53. Discuss the following example of motion in a vertical direction: —

A movable chamber 4·9 m. high can be made, by machinery, to descend the vertical shaft of a mine with the uniform velocity of 9·8 m. per second. A ball is dropped from the top of the chamber, (1) when the chamber is at rest, (2) when the chamber is descending with the uniform velocity of 9·8 m. per second.

54. If a stone be dropped from the top of a cliff, what velocity will it acquire under the action of gravity in one second? in two seconds? in t seconds? in one quarter of a second?

55. Explain with precision the statement that "at the end of one second a body falling freely will attain a velocity of 9·8 m. per second."

56. What may be called the *average* or *mean* velocity of a falling body during the first second? during the first two seconds? during t seconds?

57. Prove that, in uniform motion, space passed over is equal to velocity multiplied by time.

58. Show that any case of uniform motion may be represented graphically by the area of a rectangle.

59. Prove, by dividing a second into tenths, and supposing the motion uniform during each tenth of the second, that the space passed over by a body falling freely in the first second of its motion is 4·9 metres.

60. In general, what represents the space described by a body falling freely for any given time?

61. If t = the whole time of fall, and s = the space passed over, show, from what has already been established, that $s = 4 \cdot 9 \, t^2$.

62. Comparing the results in questions **56** and **59** it appears that the space passed over in the first second of motion is equal numerically to the mean velocity during that second. Accepting this relation as true in general (which is the fact), find the space described during the second second of motion; during the third second.

63. Suppose a projectile, as a bomb-shell, to be fired obliquely into the air; prove that its actual path under the action of gravity will be a curve, bending farther and farther from the original line of impulse. What may this curve be shown to be?

Lesson IV. — *Second Law of Motion (continued).*

64. Suppose a piece of iron to fall by the action of gravity, and also to be acted upon by a magnet so placed as to give it in one second a velocity in the same direction as gravity of 9·8 m. per second ; find the velocity acquired and the space described in one second.

65. From the results in question **64**, what may be inferred to be the proper measure of different forces applied to the same body ?

66. What have we found to be the measure of forces which generate the same velocity in bodies having different masses ?

67. In general, what product represents the magnitude of, or is the *measure* of, a force ?

68. What product measures the *momentum* of a moving body ? Define the measure of a force in terms of the momentum which it will generate.

69. Give an instance of two forces acting in different directions simultaneously on a body at the same point, and determine the path which the body will take under the joint action of the two forces.

70. Explain how a straight line may be employed to represent the point of application, the direction, and the magnitude of a force.

71. What are the two chief results respecting the second law of motion reached, one in the last Lesson, the other in the present Lesson ?

72. Give examples of forces which act in such a way as to compel a body to remain at rest.

73. What really happens when a heavy body rests on the floor ?

74. What is the effect of a force which acts on a body without changing its state of rest or motion called ?

75. When must two pressures, or statical forces, be considered equal to each other ?

76. A man in a carriage supports a half-hundred weight in his hand. The carriage and all that it contains is now in the act of falling over a precipice. Will he still continue to feel the strain of the weight upon his arm?

77. A weight equal to 100 kilogrammes rests upon a support, the weight of which support may be neglected. This support is not altogether prevented from falling, but, in virtue of the machinery with which it is connected, it is only allowed to acquire a velocity of 4·9 metres in one second. What will be the pressure on the support?

78. What two ways are there of viewing the action of two forces acting simultaneously on a body in different directions?

Lesson V. — *Forces Statically Considered.*

79. What is the "parallelogram of forces"?
80. Give a definition of the *resultant* of two forces acting along lines which intersect one another.
81. Distinguish between the resultant of two forces and what may be called their *balancing force*.
82. If two forces, $P = 6$, $Q = 8$, act on a point at right angles to each other, find the numerical value of their resultant.
83. Describe an experimental method of demonstrating the truth of the parallelogram of forces.
84. Suppose that we have two parallel forces in the form of weights acting at the ends of a straight horizontal rigid bar or *lever*, the whole resting on a fixed point or *fulcrum* between the forces. Neglecting the weight of the bar, under what condition will the system be in equilibrium? What will be the pressure on the fulcrum?
85. Define the *moment* of a force with respect to a point.
86. State the condition of equilibrium of a lever in the language of moments.
87. Give examples of levers where a comparatively small force at a great leverage produces a very great effect. What do you mean by a "great leverage"?

88. On a lever, the perpendicular distances of the lines of action of the forces from the fulcrum are often called the *arms* of the lever. What relation must exist between the lengths of the arms in order that a given force applied at the end of one arm may overcome a greater force applied at the end of the other arm ?

89. What is the condition of equilibrium of any number of forces applied in one plane to a body which is supported on a fixed point or fulcrum ?

90. On a straight lever without weight we have on the right hand of the fulcrum two forces, namely, 8 grammes at a distance of 6 centimetres, and 12 grammes at a distance of 8 centimetres ; while on the left hand we have 10 grammes at a distance of 10 centimetres. Which arm will tend to fall ?

LESSON VI. — *Third Law of Motion.*

91. In the third law of motion what is asserted of any force which alters the state of rest or motion of a body *as a whole ?* Give an illustration.

92. What does the third law of motion assert of the momenta generated *in the parts of a body or system of bodies* by the action of internal forces ? Illustrate this truth by the example of firing a gun.

93. How is the third law of motion sometimes stated ?

94. Illustrate this law by the example of a stone falling to the ground.

95. How does the discharge of a cannon which is firmly fixed to the ground furnish another illustration of the same law ?

96. According to this law of motion, what must take place whenever a man leaps upward from the ground ?

97. Suppose a bomb-shell flying along with a velocity of 200 m. per second explodes into two parts of equal weight, one of which is propelled forwards in the exact direction in which the shell is moving with an additional

velocity of 200 m. per second. Show, by means of the third law of motion, that the other half of the shell will be brought to rest in consequence of the explosion.
98. Explain the Eolipyle.
99. Explain the ascent of a rocket.

CHAPTER II.

THE FORCES OF NATURE.

LESSON VII. — *Universal Gravitation.*

100. Into what three groups may the forces of nature be divided?

101. What is the distinction between molecular and atomic forces?

102. Illustrate the general fact that some of the forces connected with molecules and atoms may be characterized as permanent while others are temporary and evanescent.

103. What is the most important and best understood force belonging to matter?

104. What question respecting terrestrial gravity did Newton ask himself, and what answer did he find by experiment?

105. What opinion on this subject was held by the followers of Aristotle?

106. How did Galileo overturn the Aristotelian dogma?

107. Describe the "guinea and feather" experiment, stating clearly what it proves.

108. What prevents us from making exact experiments on bodies falling freely?

109. What effect would changes in the force of gravity have on the oscillations of a pendulum?

110. Describe Newton's pendulum experiments, and show that they prove that the weight of a body is directly proportional to its mass.

111. Compare gravity with magnetism, as regards the relation between the acting force and the mass acted upon.

112. Show that the measure of the force of gravity which acts on one gramme is equal to 9·8. What will it be on 5 grammes?

113. How is the *vertical* direction defined ? How found by experiment ?

114. Why are plumb lines not strictly parallel ? What change in the direction of a plumb line is produced by travelling one mile on the earth's surface ?

115. What effect on the weight of a body would be produced by a change in the mass of the earth or *attracting* body ?

116. State the law of "inverse squares," or law which expresses the mathematical relation between the distance of two bodies from each other and the force of attraction between them.

117. Prove the law of "inverse squares" by the Newtonian method of comparing the force of the earth's attraction at the moon with the same at the earth's surface.

118. Give the complete statement of the law of universal gravitation.

119. Illustrate this law by supposing different numerical values for the attracting masses and their distance from each other.

Lesson VIII. — *Atwood's Machine.*

120. What is the object of Atwood's Machine ? Describe its chief parts.

121. Describe *Experiment A*, and state what it proves.

122. Describe *Experiment B*. Compare the results in experiments A and B, and state the law which they establish.

123. Describe *Experiment C*. What may be concluded, (1) by comparing together experiments A and C, (2) by comparing together the results in all three experiments, A, B, and C ?

124. Describe *Experiment D*.

125. Describe *Experiment E*. What truth do experiments D and E illustrate ?

126. Describe *Experiment F*. What is "the law of velocities" which is demonstrated by this experiment ?

127. Describe *Experiment G*, and give the "law of spaces" which it proves.

128. Show that, if a body be projected vertically upwards, the height attained is proportional to the square of the velocity of projection.

129. What is the relation in theory between the velocity of projection of a stone, and the velocity with which it strikes the ground on its return? How is this truth illustrated by *Experiment H*?

130. Neglecting the weight of the pulley in Atwood's Machine, let the one box weigh 600 grammes, and the other 400; what will be the tension of the string during the downward motion of the heavier box?

131. A body is projected vertically upwards with a velocity equal to 19·6 metres per second; what will be its velocity after it has risen 14·7 metres?

132. Give a brief recapitulation of the facts connected with the action of gravity at the earth's surface.

Lesson IX. — *Centre of Gravity, etc.*

133. Show how the force which gravity exerts on a body may be resolved into a system of parallel forces, and from this point of view give a definition of the *centre of gravity* of a body.

134. Describe a simple practical way of finding the centre of gravity of a body.

135. If we have a heavy solid resting on a base, what condition must be fulfilled in order that it may remain at rest? Prove the necessity of this condition.

136. Define *stable* equilibrium and *unstable* equilibrium, and give examples of each.

137. State a simple law which will always decide whether an equilibrium is stable or unstable. What grounds are given for the truth of this law? Illustrate its application by the example of the egg.

138. Define *neutral* equilibrium, and give an example.

139. A cone is placed on its apex on a flat horizontal surface; determine the kind of equilibrium.

140. A uniformly heavy circular wooden disk has a piece of its substance taken out, and a piece of lead inserted instead. In what position will it rest on a flat horizontal surface?

141. How will a man rising in a boat affect its stability?

142. Why is a cart loaded with hay more liable to be overturned from irregularities in the road than one loaded with the same weight of lead?

143. Describe briefly the *balance*, and show that a sensitive balance enables us to ascertain with great exactness the weight of a body.

144. Can you determine what must be the position of the centre of gravity of a balance relatively to the centre of suspension in order that the balance may be very delicate?

145. Explain the use of the pendulum, (1) in detecting changes in the force of gravity, (2) in regulating clocks.

146. What is meant by the *isochronism* of a pendulum, and how was it first discovered? What is the length of a *seconds* pendulum?

147. What is the law which expresses the relation between the time of oscillation of a pendulum and its length?

Lesson X. — *Forces exhibited in Solids.*

148. Describe briefly, and illustrate, the chief attractive forces which are exhibited in bodies.

149. Describe briefly, and exemplify, the following *resistances to deformation* which are called into action by various forces tending to alter the shape of a solid body:

(1) Resistance to linear extension.
(2) Resistance to linear compression.
(3) Resistance to cubical compression.
(4) Resistance to torsion.
(5) Resistance to flexure.

[LESS. X.] ELEMENTARY PHYSICS. 15

Which one of these resistances is also exhibited by liquids and gases?

150. Explain what *friction* is, and define the *coefficient of friction*.

151. Prove that if the pressure remain the same, the friction is independent of the magnitude of the surface.

152. Give Rennie's laws upon friction. How may friction be reduced to a minimum?

153. What conditions are favorable to the formation of crystals? Give an instance.

154. Give instances of crystals of great value which have not yet been formed artificially.

155. What conclusion about crystals is drawn from the behavior of many crystals, — for example, those of Iceland spar?

156. What peculiarities of *structure* and what consequent properties are exemplified by such substances as wood and flax? by wrought-iron? by such substances as mica and oyster-shells?

157. Give examples of solids exhibiting no apparent trace of structure. In general, what effect does time and vibration have on the structure of a solid? Give an example.

158. Define *tenacity*. How are experiments on tenacity conducted? What law is established by the experiments? What convenient measure of tenacity does this law point out?

159. What do the results of Wertheim's experiments, as given in the table, show as to the effect of *time* on tenacity?

160. What peculiarity is there in the tenacity of fibrous solids like wood? According to Musschenbroeck, which is the most tenacious of woods?

161. Instead of bodies suddenly giving way under forces tending to pull their particles asunder, what other form of rupture is common?

162. What does *ductility* denote? Give examples of bodies which possess it?

163. What is *malleability*? Which metal is the most malleable?

164. Explain *brittleness*. In what sense is a sheet of glass stronger, and in what sense is it weaker, than a sheet of paper?

165. Define *hardness*. If we have three bodies A, B, and C, how are their proper places on a scale of relative hardness found?

166. Which is the hardest of all known substances, and how is it cut?

167. Explain the processes known as *tempering* and *annealing*. What are Prince Rupert's drops?

168. What does the word *elasticity* denote? What is meant by the *limit of perfect elasticity*?

169. What condition should a solid structure such as a bridge fulfil?

170. Give the laws which hold true, within the limit of perfect recovery, of resistance to linear extension.

171. What relation exists between the forces which will produce equal amounts of linear extension and linear compression respectively in the same body?

172. How are the forces with which different substances resist linear extension or compression compared together?

173. How are experiments on torsion conducted? What laws have been established?

174. Mention some of the ways in which the force with which a solid resists any attempt to bend it is utilized.

175. What are the laws of flexure? What follows from the last law as to the best form for a beam of given mass which is to be heavily loaded?

Lesson XI. — *Forces exhibited in Liquids.*

176. What is the essential difference between a solid and a liquid? What are proofs that cohesion is not entirely wanting in liquids?

177. Describe the state of liquidity called *viscous*. Give instances to show that time is an element of importance in determining the liquidity of a substance.

178. What is characteristic of the resistance to compression offered by liquids? What is the exact measure of the compressibility of water?

179. State the law of liquid pressure discovered by Pascal, and illustrate it by the imaginary case of a hollow vessel full of water which is uninfluenced by gravity.

180. Show that, by employing pistons of different sizes, a fluid is capable of forming a very powerful mechanical arrangement.

181. What is the principle, and what are some of the uses, of Bramah's press?

182. Prove that the surface of a liquid in an open vessel must be perpendicular to the force of gravity, that is, *horizontal*. What is the true character of the surface of a large body of water, as the ocean?

183. Explain the construction and use of the *water-level*.

184. Explain the construction and use of the *spirit-level*.

185. Explain Artesian wells.

186. What is the measure of the pressure on the horizontal base of an open vessel full of a liquid? Hence show what must be the relation between the pressure on any horizontal layer of liquid in an open vessel and (1) the depth of the layer, (2) the area of the layer.

187. Show by a simple experiment that the pressure of a horizontal layer of water is the same upwards as downwards.

188. A hollow cubic decimetre, open at the top, is filled with water; what will be the pressure on the bottom and sides?

189. A vessel contains water to the depth of a decimetre, and one of the sides of this vessel is a rectangular surface, the bottom of which is one decimetre, while the

B

side slopes at an angle of 45°; what is the whole pressure on this side ?

190. In what way is the pressure exerted by a liquid connected with the *density* of the liquid ?

191. Prove that a fluid buoys up a solid immersed in it with a force equal to the weight of the fluid displaced.

192. Apply the principle of buoyancy successively to the cases in which the density of the solid immersed is greater than, equal to, and less than that of the fluid.

193. A cube of wood, the density of which is equal to 0·8, is put into a vessel containing water; what portion of its side will be immersed ?

194. Taking water at 4° C. as the standard of specific gravities or relative densities, how do you define the specific gravity of any substance ? If, as in the Metric system, the density of water at 4° C. is equal to unity, what relation must exist between the specific gravity and the density of any substance ?

195. Explain a method of finding the specific gravity of a solid body. Suppose, for the sake of illustration, that a substance weighs in vacuo 120 grammes, and when immersed under water at 4° C. only 89 grammes; find the specific gravity of the substance. Hence, what general rule may be given for obtaining the specific gravity of a solid substance ?

196. Explain a method of ascertaining the specific gravity of a liquid. As an illustrative example, let the loss of weight of a solid body in water equal 31 grammes, and its loss of weight in the liquid in question equal 28 grammes. Find the specific gravity of the liquid.

197. Describe some capillary phenomena which show; (1) what two kinds of capillary action exist, (2) what influence the diameter of the tube has on capillary ascent and depression.

198. Mention other illustrations of the laws of capillarity.

199. Describe *endosmose* and *exosmose*.

Lesson XII. — *Forces exhibited in Gases.*

200. In what respect does a gas differ from a liquid? In what respects is it like all other substances?

201. Describe an experiment illustrating that a gas has weight, and also that some gases weigh more than others.

202. How may a liquid be converted into a gas?

203. True steam being invisible, how do you account for the visible cloud arising from a kettle or a railway engine?

204. State and illustrate the distinction between *gases* and *vapors*.

205. What agencies tend to bring a gas into the liquid or solid state? What six gases have never yet been liquefied? What substance has never yet been vaporized?

206. What is the composition of the atmosphere?

207. What effect on the air would the processes of respiration and combustion, if unbalanced, in course of time produce? Why does the composition of the atmosphere in point of fact remain unchanged?

208. Why is it that, although the air is exerting pressure all around us, we seldom perceive any traces of it?

209. What is the experiment of the Magdeburg hemispheres, and what does it illustrate?

210. How was the ascent of water in pumps accounted for up to the time of Galileo? Give Torricelli's reasoning, together with his celebrated and decisive experiment, upon this subject.

211. How was the truth of Torricelli's discovery verified by Pascal?

212. What remarks are made by the author in regard to the connection between the barometer and the weather?

213. If we cork a flask full of air and then remove half of the mass of air within the flask, how much will the pressure on the interior of the flask be changed? What is that statement of Boyle's law, which gives directly the relation between the *mass* and the *pressure* of a quantity of gas?

214. State the law of Boyle in the form adopted by the discoverer, and verify its truth by a simple experiment.

215. Show that the two forms of stating Boyle's law come to the same thing.

216. What is the true explanation of gaseous pressure according to many philosophers? Show, with the aid of a numerical illustration, that their hypothesis is in harmony with Boyle's law.

217. Show, by an experiment, that gases as well as liquids possess buoyancy, and are subject to the principle of Archimedes.

218. What is the reason that a *balloon* rises in the atmosphere?

219. Describe the construction and action of an *air-pump*.

220. Prove that the density of the air in the receiver diminishes in a geometrical ratio, *i. e.* that a constant fractional part of the mass of air remaining in the receiver, is expelled by each double stroke. Why could we never, even in theory, succeed in producing a perfect vacuum? When is a practical limit reached?

221. Describe the construction and action of the common *lifting-pump*. What sets a limit to the height to which water can be raised by means of this pump?

222. Describe a *siphon*, and its action, and explain why the flow of liquid from one vessel to the other is maintained. When will a siphon once set in action cease to act? What fixes a limit to the working length of the shorter arm of a siphon?

223. Describe Graham's experiment on gaseous diffusion.

224. Mention substances which have the power of absorbing or retaining matter in the gaseous state.

CHAPTER III.

ENERGY.

Lesson XIII. — *Definition of Energy.*

225. Adduce examples which illustrate the application, up to recent times, of Newton's law of action and reaction. What view did the old hypothesis take of the phenomena of collision and friction?

226. Mention the considerations which probably led the way to a numerical estimate of work.

227. Define the unit of work, and show how to find the amount of work done in lifting a body to any given height.

228. Define energy, and estimate how much energy will be imparted to a stone weighing one kilogramme by projecting it vertically upwards with a velocity of 9·8 metres per second.

229. Prove that the work which can be accomplished by a moving body is proportional, (1) to the square of its velocity, (2) to its mass.

230. Show that the work capable of being done by a body whose mass (in kilogrammes) is m, and whose velocity (in metres) is v, is represented by the expression $\frac{m v^2}{19.6}$. (The general form is $\frac{m v^2}{2 g}$.)

231. Estimate the energy of a body weighing 64 grammes projected vertically upwards with the velocity of 60 metres per second.

232. Show, by means of the force of gravitation, that two types or kinds of energy exist, which are mutually convertible.

Lesson XIV. — *Varieties of Energy.*

233. Give a summary of what was said about the two forms of energy in the last Lesson.

234. Illustrate the transformation of energy by an example, taken from chemistry, and point out the analogies to the case of gravitation. What appears from this example to be the true nature of heat?

235. Name some of the varieties of visible or mechanical energy, both kinetic and potential.

236. How does the doctrine that heat is a form of energy explain the phenomena of latent heat?

237. Trace the analogy between the mechanical world and the molecular world by means of the phenomena of sound and of radiant light and heat.

238. Show that electrical phenomena afford another illustration of molecular energy.

239. What great advantage as a source of energy does electricity in motion possess over water in motion or heat?

240. Give a brief recapitulation of the various forms of energy.

Lesson XV. — *Conservation of Energy.*

241. What was formerly in many minds the great ideal of mechanical triumphs?

242. What conclusive answer may now be made to the arguments in favor of perpetual motion?

243. What is the principle of the conservation of energy, and what is the nature of the evidence in favor of its truth?

244. Apply the principle of the conservation of energy to the case of a stone projected vertically upwards.

245. What becomes of the energy of a railway train when it is suddenly stopped? or of a cannon-ball after it has struck the target? or, in general, what becomes of the energy of visible motion when it has been stopped by percussion or friction?

246. What experimental evidence upon this point do we owe to Rumford and Davy? What simple phenomena can you mention which furnish evidence in the same direction?

247. Who were the first to point out the probability of a connection between the various forms of energy? Who established this connection on a scientific basis? In particular, what was the result of the researches of Joule?

248. Illustrate still further the connection between the various kinds of energy by what takes place in a galvanic battery.

249. Give an algebraic statement of the doctrine of the conservation of energy.

250. Illustrate the true function of a machine, by applying the law of the conservation of energy to one of the ordinary mechanical combinations, such as a system of pulleys.

251. Apply the same law to the hydraulic press.

252. What law holds universally in machines, if friction be left out of account, respecting the power, the weight, and the distances which they traverse?

CHAPTER IV.

VISIBLE ENERGY AND ITS TRANSMUTATIONS.

LESSON XVI. — *Varieties of Visible Energy.*

253. Mention some of the varieties of visible kinetic energy.

254. Why is the energy of a vibrating string classed among the forms of visible energy?

255. Give examples of visible potential energy.

256. Suppose a rifle-ball weighing 20 grammes and moving with a velocity of 200 metres per second buries itself in a block of wood weighing 20 kilogrammes and hung by a string so as to form a pendulum. Find how much of the energy of the ball will reappear after the impact in the form of visible energy. What has become of the remainder of the energy of the ball?

257. What distribution of momentum, and what consequent change of energy, take place during the passage of a ball through the air?

258. What extension of the first law of motion are we now prepared to recognize?

259. Examine the following case of direct impact of two *inelastic* solids, and find what part of the united energy of the two masses before impact is transmuted into heat: — weights of the solids, 20 grammes and 10 grammes; velocity of the first, 20; of the second, 16 in an opposite direction.

260. What is the law of energy which applies to the case of the impact of two *perfectly elastic* bodies? Suppose, as an illustrative example, that two perfectly elastic balls, weighing respectively 4 and 3 kilogrammes, moving in the same direction with velocities 5 and 4, impinge on each other.

261. Apply the laws of impact to the case in which one elastic ball impinges directly against the extremity of a row of elastic balls at rest.

262. Consider briefly the energy of a circular disk in rapid rotation. In general, wherever in nature a body revolves in a circle round a central force, what is true of its velocity and of its kinetic energy?

263. Consider the energy of a body moving in an ellipse, and show that a definite part of its energy assumes alternately the kinetic and the potential forms.

264. Show that the doctrine of energy leads to the conclusion that the velocity of a body which has slid down a smooth inclined plane depends only on the height of the plane. Why does this proposition fail in case the plane be rough?

265. Mention some instances of visible energy of position, and state how the energy in each case may be converted into energy of motion.

266. Show that the energy of an oscillating pendulum is alternately kinetic and potential.

267. A pendulum bob weighing one kilogramme is so swung that it is higher at the summit of its oscillation than at its lowest point by one decimetre; what is its velocity at its lowest point?

268. Describe Foucault's pendulum experiment, the pendulum being supposed, for the sake of simplicity, to oscillate at the north pole.

269. Consider the energy of a vibrating body, as the string of a musical instrument or a bell, and point out the analogies between the motion of such a body and that of a pendulum.

270. What effect does vibratory motion in the air usually produce before it assumes the shape of heat?

271. Give a recapitulation of the various kinds of visible energy which have been considered in this Lesson.

Lesson XVII. — *Undulations.*

272. Prove that for small vibrations of a pendulum the force which urges the ball in the direction of its motion at each instant is proportional to the distance of the ball from its point of rest, *i. e.* is proportional to the *displacement.*

273. Explain the principle of *isochronism.* How far is it applicable?

274. Upon what elements does the time of vibration of a body depend? How may this be illustrated?

275. Define *wave motion*, and give examples of it.

276. Define an *up-and-down* or *transverse* wave, and a *wave-length;* and illustrate by diagrams the manner in which an undulatory motion is propagated by an up-and-down wave.

277. If v denote the velocity with which the wave is propagated, t the time of a double vibration of a particle, and l a wave-length, show that $l = v\,t$.

278. What is the nature of waves *of condensation and rarefaction,* or *longitudinal* waves?

279. What is meant by the *phase* of a vibrating particle?

280. What is an essential peculiarity of vibrating motion as regards the phases of any two contiguous particles?

281. What is the *amplitude* of a vibration?

282. Explain why the wave-length does not depend on the amplitude of the vibration.

Lesson XVIII. — *Sound.*

283. What is the definition of Acoustics?

284. In what two senses is the word "sound" used? In which sense is it here used?

285. Explain the difference between a *noise* and a *musical sound* or *note.*

286. What determines the *pitch* of a note?

287. What other characteristics of sound besides pitch are perceived by the ear?

288. What is the nature of sound-waves?

289. How may it be shown that sound is not propagated in vacuo?

290. What are the laws of the reflection of sound? Illustrate them by means of a diagram.

291. What constitutes an echo? When, for instance, is an echo more audible than the original sound?

292. What condition must be fulfilled in order to produce a distinct echo? What effect do whispering-galleries have upon a sound?

293. What experiment upon sound may be performed with two conjugate reflectors?

294. What is related of the Cathedral of Girgenti in Sicily?

295. What power does a convex glass lens exercise upon rays of light?

296. What was the experiment of M. Sondhauss, and the conclusion to be drawn from it?

297. What is the velocity of sound in air, and by what method was it determined?

298. What effect, if any, have pitch and intensity respectively on the velocity of sound?

299. How does the velocity of sound in air compare with its velocity in hydrogen gas at the same pressure? What reason can be given for this difference?

300. Explain why the velocity of sound in a given medium does not vary with its density.

301. Why does sound travel faster in warm than in cold air?

302. What is the velocity of sound in water? in wood?

303. Prove that the intensity of a sound varies inversely as the square of the distance from the source.

304. Why is it probable that the intensity of sound really diminishes somewhat more rapidly than is indicated by the law of "inverse squares"?

305. What effect does a change in the density of the

medium have upon the intensity of sound? How may this law be verified? What is the reason that it is true?

306. Why is sound much better heard when the air is calm and homogeneous?

307. By what arrangement may the intensity of the sound of a musical string be strengthened?

LESSON XIX. — *Vibrations of Sounding Bodies.*

308. State the laws which connect the vibrations of a stretched string with the properties of the string.

309. Explain the mode of action of an organ pipe.

310. What is the reason that an organ pipe filled with any other gas than air yields an entirely different sound? What use may be made of this fact?

311. What acoustical difference is there between a shut pipe and an open pipe of the same dimensions?

312. Suppose we have rods of wood fixed at one end and free to move at the other; what two kinds of vibratory motion may be imparted to them?

313. What are the laws which express the relations between the number of vibrations of a rod and its dimensions?

314. What is the law which governs the vibrations of plates?

315. What are *nodal points* or *nodes* on a vibratory cord, and how may they be produced? What is a *loop* or *ventral segment?*

316. In producing nodes along a cord, what condition must be observed in order that the vibrations shall not interfere with each other?

317. How may the existence of nodes and ventral segments be rendered evident by means of vibratory plates?

318. What law governs the communication of vibrations from one instrument to another through the air? Mention instances in which this communication may be observed.

319. Explain the action of Savart's machine for measuring the number of vibrations corresponding to a given sound.

320. Describe Lissajous's method of making vibrations apparent to the eye.

CHAPTER V.

HEAT.

LESSON XX. — *Temperature.*

321. What two kinds of energy are denoted by the word heat?

322. What three-fold division of the study of phenomena connected with heat will be adopted in this chapter?

323. When are two bodies said to be of the same temperature? When is one body said to be of a higher temperature than another? How would you define temperature?

324. What is the general law of expansion under the action of heat? Cite experiments which go to prove this law.

325. What is one remarkable exception to the above law of expansion?

326. Why in general does expansion fail to furnish us with an exact measure of temperature? How does the case of water illustrate this?

327. Why is a substance near the point of changing its state not fitted to be used as a means of measuring temperature?

328. Compare the relative merits of mercurial and air thermometers as regards accuracy and convenience.

329. What is the principle of the mercurial thermometer? Describe the process of filling the bulb and tube with mercury.

330. How may great delicacy be secured in a thermometer?

331. What are the two *fixed points* of temperature on a thermometer? Describe how they are determined.

LESS. XX.]   *ELEMENTARY PHYSICS.*   31

332. Explain how the stem of a thermometer is graduated according to the centigrade scale. How do the Fahrenheit and the Réaumur scales differ from the Centigrade?

333. Deduce formulæ for reducing Centigrade degrees to Fahrenheit, and *vice versa*.

334. Find the degree of Fahrenheit which corresponds to 45° Centigrade.

335. Find the degree Centigrade which corresponds to 86° Fahrenheit.

336. What degree Fahrenheit corresponds to 40° Centigrade?

337. Deduce formulæ for reducing Centigrade degrees to Réaumur, and *vice versa*.

338. Explain the *change of zero* of a mercurial thermometer, and the corrections which must be applied to eliminate this source of error.

339. What is the effect of suddenly heating and cooling a thermometer? What practical rule does this effect suggest with regard to the determination of the fixed points?

340. Explain why it may be necessary to apply a correction to the reading of a mercurial thermometer on account of the position in which the instrument is held.

341. Suppose a mercurial thermometer is placed under the receiver of an air-pump and the air exhausted, what will be the effect on the column of mercury?

342. If the bulb of a thermometer, together with the lower part of the stem up to the zero-point, be immersed in boiling water, while the remainder of the stem from 0° to 100° is exposed to air at the freezing-point, the temperature denoted will be 98·4° instead of 100°. Explain this.

343. Why is mercury unsuited for measuring very low temperatures? What liquid is used for this purpose?

344. Describe a *minimum* thermometer and its action.

345. Describe Phillip's *maximum* thermometer and its action.

346. Describe Leslie's *differential* thermometer and its action.

LESSON XXI.—*Expansion of Solids and Liquids through Heat.*

347. Explain (with a diagram) how the comparatively small expansion of a solid rod due to heat may be rendered visible by mechanical means. What other method of magnifying expansion is also employed?

348. Define the *coefficient of linear expansion* of a substance.

349. Which is the most expansible of the metals mentioned in the table? which the least?

350. What circumstances may modify the value of the coefficient of expansion of a substance? Give an example.

351. Show that the cubical expansion of a substance is very nearly three times as great as the linear for the same change of temperature.

352. State the principle of one method of determining experimentally the cubical expansion of solids.

353. Illustrate this method by solving the following example:—

A solid weighs 600 grammes in vacuo, and only 400 grammes in a fluid at 0° C., of which the specific gravity is 1·2, while it weighs 406 grammes in the same fluid at 100° C., for which temperature the specific gravity of the fluid is known to be 1·16; find the cubical expansion of the solid between these two temperatures.

354. What general relation between the coefficients of linear and cubical expansion of any substance is indicated by the table on page 161?

355. What notable exceptions are there to the following laws:—

(1) Solids expand through heat,
(2) And expand equally in all directions.

356. What effect, in general, has the temperature of a solid on its rate of expansion?

357. Explain the distinction between the *apparent* and the *real* expansion of a liquid.

LESS. XXII.] *ELEMENTARY PHYSICS.* 33

358. Give an outline of the method of finding the real expansion of a liquid, called the method by thermometers.

359. Explain Matthiesson's method of determining the real expansion of a liquid.

360. Illustrate Matthiesson's method by solving the following example : —
A piece of glass, of which the linear expansion from $0°$ to $100°$ C. is known to be 0.0009, loses at $0°$ C. one gramme of its weight in the fluid in which it is weighed, while at $100°$ C. it only loses 0.96 of a gramme. Find the expansion of the fluid between $0°$ and $100°$ C.

361. Give an outline of Regnault's method of determining the expansion of mercury.

362. What peculiarity does water exhibit with respect to its expansion?

363. Explain Hope's method of ascertaining the point of maximum density of water.

364. What fact respecting the rate of expansion of water at different temperatures is shown by the table on page 165?

365. What reasons do we have for believing that the rate of expansion of very volatile liquids must be very great? How is this conclusion verified in the case of liquid carbonic acid?

366. Recapitulate the chief laws for the expansion of liquids.

LESSON XXII. — *Expansion of Gases. Practical Applications.*

367. State and illustrate the law discovered by Charles, which expresses the relation between the temperature and the pressure of a gas, the volume remaining constant.

368. Prove that the same law expresses also the relation between the temperature and the volume of a gas, the pressure remaining constant.

2* C

369. A bladder which at 0° contains 900 cubic centimetres of air has its temperature increased to 30° C., the pressure under which the gas exists meanwhile remaining constant ; what will now be the volume of the gas in the bladder ?

370. What are some direct consequences of the fact that the coefficient of expansion is the same for all gases ?

371. Explain more fully than was possible in Lesson XX. the advantage of using an air thermometer.

372. At what temperatures are the metre and the yard respectively *standards* of length ?

373. Explain why it is necessary in all very accurate weighings to know the temperature of the air.

374. Show that, in the metric system, the weight (in grammes) of one cubic centimetre of any substance will denote at the same time its specific gravity.

375. Why is it necessary to fix upon a standard temperature in comparing the specific gravities of substances, and what is this standard ?

376. What is the standard pressure employed in comparing the specific gravities of gases ?

377. Explain what effect change of temperature will produce upon the motion of a pendulum, and also upon the motion of the balance-wheel of a chronometer.

378. Explain *Harrison's gridiron pendulum*.

379. Explain the *compensation balance* for chronometers.

380. Mention other instances in which account must be taken of the expansion of bodies.

381. Mention instances in which advantage is taken of the fact of expansion.

LESSON XXIII. — *Change of State and other Effects of Heat.*

382. What invariable rule holds in the production of changes of state through heat ?

383. Give examples of substances which differ from

one another in the manner of their passage from the solid to the liquid state. What is this passage called?

384. Give instances in which change of composition accompanies change of state.

385. What effect does pressure have on the melting-point of ice? What general law is found to hold true with respect to the connection between pressure and congelation?

386. What is the reason that gold, silver, and copper coins cannot be cast in a mould, but must be stamped?

387. Under what conditions may water be cooled below the freezing-point without becoming ice? What other instances of a similar phenomenon exist? In all such cases how may solidification be immediately produced?

388. What is *regelation*, and what is Forbes's explanation of it?

389. What is the difference between sublimation and vaporization?

390. Distinguish between two kinds of vaporization.

391. What did Dalton show with respect to the formation of a vapor in a confined space?

392. How far is the evaporation of a liquid modified by taking place in air instead of in vacuo?

393. What conditions are most favorable to rapid evaporation in the open air, and why?

394. Explain the apparatus and process of distillation.

395. What is ebullition, and how is it affected by the supply of heat?

396. Enumerate the chief circumstances upon which the boiling-point of a liquid depends.

397. What are the boiling-points of ether and of mercury, as given in the table on page 178?

398. Describe two simple experiments which show that the boiling-point of a liquid depends upon the pressure under which it exists.

399. Why is the boiling-point of water at the top of a mountain lower than at its bottom? How does this fact interfere with culinary operations, and how is the difficulty remedied?

400. In what way is the fact that the boiling-point of water is lessened as we rise above the sea-level applied to a practical use?

401. How do glass and metal vessels compare with each other in their influence upon the boiling-point of water? What is the effect of dropping iron-filings into the glass vessel?

402. In what way was Donny able to raise the temperature of water to 135° C. without ebullition?

403. What is the general effect of salts in solution upon the boiling-point of water?

404. Give examples of the behavior of liquids in the *spheroidal state.*

405. In what way can the behavior of liquids in the spheroidal state be explained, and what experiment of Boutigny confirms the truth of this explanation?

406. What was Faraday's experiment with ether, solid carbonic acid, and mercury, and how do you explain the effects which he observed?

407. Andrews heated liquid carbonic-acid, under great pressure in a closed tube, to the temperature of 31° C., or thereabouts. What phenomena were observed, and what inference has been drawn from them?

408. What is sublimation, and what are instances of it?

409. What is the effect of heating a strong solution of hydrochloric acid in water? of heating chalk?

410. Give an illustration of the great attraction which some gases have for water.

411. What six gases have never yet been condensed by the joint effect of cold and pressure?

412. Explain what is meant by the *maximum pressure* of a vapor.

413. Show how, by means of the table on page 183, we may obtain the atmospheric pressure by observations of the boiling-point thermometer.

414. What was the discovery of Gay Lussac in reference to the densities of gases? Illustrate it by the case of hydrogen and chlorine.
415. Give a recapitulation of the effects of heat which have already been considered.
416. In addition to the effects already discussed, enumerate other ways in which heat influences bodies.

LESSON XXIV. — *Conduction and Convection.*

417. In what way do we derive our heat from the sun?
418. Give an instance of the mode of distribution of heat called *conduction*. How does it differ from radiation?
419. What example does this branch of the science of heat afford of the provision by nature for the welfare of the animal creation?
420. For what two purposes may a bad conductor of heat be employed? Give an instance of each.
421. Give an experiment which shows the difference between the conducting power of two different substances.
422. Why is it that, when a metal bar has one of its extremities heated in the fire, the other extremity does not *ultimately* attain the same temperature?
423. Give Fourier's definition of the *conductivity* of a substance.
424. Suppose we have two bars of the same shape, size, and conductivity, but unlike in material, the ends of which we heat by a spirit-lamp to the same extent; and let the surfaces of both bars be covered with gilt. Now, if we observe the temperatures of both bars at equal distances from the lamp, we shall find them *unequal* or *equal* according as a *short time* or a *considerable interval* has elapsed since the lamp was first applied. Explain this.
425. If we take two precisely similar pieces, one of bismuth and the other of iron, and, coating one end of each with white wax, place the other end in a hot ves-

sel, we shall find that the wax will melt first on the bismuth, although iron is the best conductor. Account for this.

426. Explain the principle of Davy's safety-lamp.

427. What peculiarity do crystals exhibit in their conducting power, and how was this shown experimentally by De Senarmont?

428. How may the bad conducting power of water be shown by experiment?

429. Describe the process called *convection*. How may convection currents be rendered visible?

430. Give an account of convection on the large scale, as exemplified by the freezing of a lake. What consequences would ensue if water had no point of maximum density and ice were heavier than water?

431. Upon what two things does convection depend? How is this illustrated by the atmosphere of the sun?

432. Explain the trade-winds.

433. Explain the land and sea breezes.

LESSON XXV. — *Specific and Latent Heat.*

434. Define the *unit* of heat; and also the *specific heat* of a substance.

435. Explain the method of determining the specific heat of a substance called the "method of mixtures" by the aid of a numerical example.

436. What other methods of estimating specific heat have been devised?

437. In general, what influence have temperature and density respectively on the specific heat of solids?

438. Generally speaking, do substances have a greater specific heat in the solid or in the liquid state?

439. What substance has generally been supposed to have the greatest specific heat?

440. What two kinds of specific heat may be distinguished in the case of a gas?

441. What results did Regnault obtain respecting the specific heats of gases?

442. What law did Dulong and Petit discover with respect to the specific heats and atomic weights of simple substances?

443. Under what circumstances does heat become *latent*? How may we correctly describe the condition of water at 0° C., or of steam at 100° C.?

444. How were Black's first experiments upon latent heat performed? How were his subsequent experiments performed so as to measure the latent heat of one kilogramme of water?

445. What is the value, in the metric system, of the latent heat of water? of the latent heat of steam?

446. What important parts do the facts that water has a greater latent heat than any other substance, and that steam has a greater latent heat than any other gas, play in the economy of nature?

447. Viewing heat as a species of molecular energy, what twofold office does it discharge?

448. What explanation of the phenomona connected with latent heat is furnished by the doctrine of energy?

449. What is the principle of all freezing mixtures and processes?

450. Explain the method of using the wet and dry bulb thermometers for estimating the hygrometric state of the air.

451. How did Leslie freeze water by means of its own evaporation?

452. Describe Carré's apparatus for artificially producing ice?

453. How did Faraday succeed in freezing mercury?

454. Explain the fall of temperature which results from mixing snow and salt together.

LESSON XXVI. — *On the Relation between Heat and Mechanical Energy.*

455. Give instances of the conversion of mechanical energy into heat. In what case is this conversion undesirable, and what means are taken to avoid it?

456. How did Joule conduct his experiments on the relation between mechanical energy and heat?

457. What is the mechanical equivalent of heat as determined by Joule?

458. What was the nature of Mayer's method of calculating the mechanical equivalent of heat?

459. If we drop a weight into a large quantity of fulminating powder, the result is the generation of a large amount of heat; are we at liberty to suppose that all this heat is the mechanical equivalent of the energy of the weight?

460. What question similar to the above may be asked in the case of the compression of a gas, and what answer to it is supplied by Joule's experiments?

461. Give the mechanical explanation of the fact that a gas suddenly expanded becomes cooled.

462. Explain, with the aid of a diagram, how the alternating motion of the piston in the cylinder of a steam-engine is produced through the agency of steam.

463. What are the chief differences between *high-pressure* and *low-pressure* engines?

464. What is the general law for the conversion of heat into mechanical energy? How is this law exemplified by low-pressure and high-pressure engines respectively?

465. State Carnot's analogy between the mechanical capability of heat and that of water?

466. What is the absolute zero of temperature, and its value on the centigrade scale?

467. Under what conditions would it be possible to convert all the heat which passes through a heat-engine into mechanical effect?

468. Suppose that the higher temperature of a heat-engine is 100° C., and the lower 0° C.; what proportion of the whole heat carried through the engine may be converted into mechanical effect?

469. What is the general rule for finding how much of the heat carried through a heat-engine can be utilized?

470. What difference in general is there between the theoretical and the practical limit of utilization?

471. What three kinds of heat-engines are in extensive use?

472. Give a sketch of the history of heat-engines prior to the time of Watt.

473. How was Watt's attention drawn to the subject, and what were the three great improvements which he effected?

474. Explain the advantages of Watt's arrangement for the *condensation* of the steam.

475. Explain the principle of *double action* introduced by Watt.

476. Explain the mode of *expansive working*.

477. How is the rate at which an engine performs work usually expressed, and what is the unit employed for this purpose called? What is its numerical value in this country?

478. On reviewing the relations of heat and mechanical effect, what important difference in their mutual convertibility is apparent?

CHAPTER VI.

RADIANT ENERGY.

LESSON XXVII. — *Preliminary.*

479. What is the velocity with which radiant energy is propagated ?

480. Mention phenomena which naturally lead to a division of radiation into *non-luminous* and *luminous,* or into rays of *dark heat* and rays of *light.*

481. Define *Optics.*

482. What are the two hypotheses respecting the nature of light ? Why has the new hypothesis the better claim to be regarded as true ?

483. Define the following terms : a *ray* of light, a *pencil* of rays, *divergent* pencil, *convergent* pencil, pencil of *parallel* rays.

484. Into what two distinct classes are substances divided with reference to their effect upon light ?

485. What is the effect of allowing light to fall on a very thin slice of an opaque substance ? What does this show as to the distinction between opaque and transparent substances ?

486. What two important exceptions are there to the law that light moves in straight lines ?

487. Explain (with a diagram) how Römer was able to determine the velocity of light from the eclipses of Jupiter's satellites.

488. Explain Fizeau's method of measuring the velocity of light.

489. Prove that the quantity of light which a surface receives from any source will vary inversely as the square of its distance from the source.

490. Show that the intensity of the illumination of a plate or screen is proportional to the cross-section which it presents to the direction of radiation.

What familiar facts respecting the power of the sun's rays are accounted for by this law?

491. Prove that the intrinsic brightness of a luminous body does not vary with its distance; meaning by brightness the light that would reach the eye by looking at the body through a long narrow tube, and supposing the tube to be always so narrow, and the source of light always so large, that in looking through the tube we should see nothing else but this light.

Consider also the case in which the luminous body is so distant as to appear simply a luminous point like a star.

492. Explain *Bunsen's photometer*, and the method of measuring the intensity of light by means of it.

493. Suppose that one light causes the grease spot to vanish in a Bunsen's photometer when placed at the distance of one foot in front of the screen, and another light when placed at the distance of two feet; what is the relative luminosity of the two lights?

494. Explain the distinction between the illuminating power of a source of light and the inherent brightness or quality of the light.

Lesson XXVIII. — *Reflection of Light.*

495. State the law of reflection.

496. How may the truth of this law be rendered visible to the eye?

497. What is meant by a *virtual* image?

498. Prove by the aid of a figure that the image of a luminous point lies as far behind the reflecting surface as the luminous point itself lies before it.

499. Explain by a figure the mode of determining the positions of the various points in the image of a luminous body which is in front of a plane mirror.

500. What peculiar inversion is there in the reflection of the human figure in a vertical mirror? Also, in the reflection of letters written from *left to right* on a wall in front of the mirror?

501. When a ray of light strikes a curved surface, how may the direction of the reflected ray be found?

502. Suppose a pencil of parallel rays strikes a concave spherical mirror; prove that the focus of the rays is the point half-way between the centre of the mirror and the middle point of its surface. What is this focus called? (See page 229.)

503. Explain how a concave spherical mirror produces a circular image of the sun. Will this image be *real* or *virtual*?

504. Show by a diagram that the focus of divergent rays proceeding from a point near a concave spherical mirror lies between the principal focus and the centre of the mirror.

505. What are *conjugate foci*, and what is meant by saying that conjugate foci are interchangeable?

506. Prove the formula which gives the relation between the conjugate foci of a concave spherical mirror.

507. Show that a virtual image will be produced if the luminous point be nearer the concave mirror than the principal focus.

508. Examine the five different cases to which the general formula $\frac{1}{d}+\frac{1}{D}=\frac{2}{r}$ is applicable.

509. The images produced by concave mirrors are in their nature either *real* or *virtual*, in their position relatively to the object either *erect* or *inverted*, in size compared with the object either *magnified* or *diminished*. Examine as regards these particulars the image of an object such as a straight line placed *beyond* the centre of the mirror.

510. Examine as regards the above particulars the image of an object placed between the principal focus and the concave mirror.

511. An object is placed immediately in front of a concave mirror, and then gradually removed to a great distance along the axis of the mirror; trace the changes in the *distance* of the image from the mirror, and also in the *nature, position,* and *size* of the image.

512. Explain, with a diagram, the effect produced by a parabolic mirror upon a pencil of parallel rays incident on its surface. What advantage do parabolic mirrors possess as compared with spherical mirrors? On the other hand, what disadvantage?

513. What is the nature of the images produced by convex spherical mirrors?

Lesson XXIX. — *Refraction of Light.*

514. Illustrate, by a diagram, the refraction of light by the surface of a transparent medium like glass, and state the law of refraction in general terms.

515. Instead of introducing the sines of the angles of incidence and refraction in the statement of the law of refraction, how can the law be expressed in purely geometrical language?

516. Explain the case in which the ray of light, instead of passing from vacuo into a transparent medium, passes out from the medium into vacuo.

517. Explain the case in which a ray of light strikes the surface of a medium at right angles.

518. Explain how the truth of the laws of refraction may be illustrated experimentally.

519. Explain *total internal reflection.* What is the *critical angle* of a medium?

520. Explain the *mirage.*

521. If n_1, n_2 be the absolute indices of refraction of two media respectively, and n' be the relative index of refraction for the two media, prove that $n' = \frac{n_2}{n_1}$.

522. Trace, by means of a figure, the path of a ray of

light through a glass prism. What is the *angle of deviation* ?

523. What is the condition of *minimum deviation* in a prism ?

524. What is the condition of *total internal reflection* in a prism ?

525. Why can we not employ for ordinary optical purposes a glass prism, of which the angle is greater than 84° ?

Lesson XXX. — *Lenses and other Optical Instruments.*

526. Describe the shapes given to the lenses in common use, and give the names of the lenses

527. From the action of a prism on a ray of light derive a rule for determining whether a lens is converging or diverging ; and apply this rule to the six lenses mentioned in the book.

528. Discuss the formula which expresses the relation between the conjugate foci of a double convex lens, examining the different cases which arise as the luminous point is supposed to move along the axis of the lens from an infinite distance up to the lens. The formula is $\frac{1}{p} + \frac{1}{p'} = \frac{1}{f}$, in which p, p' are the conjugate foci, and f the principal focus of the lens.

529. Show, by a figure, how to find the position and size of the image of a luminous body formed by a double convex lens, the luminous body being supposed to be farther from the lens than the principal focus. What is the law which determines the size of the image compared with that of the object ?

530. Show that if a luminous body be placed between a double convex lens and its principal focus, the image will be virtual, erect, and magnified.

531. Describe the two sets of appearances which may be seen in looking through a double convex lens, and state the conditions under which they are produced.

532. Describe the *camera obscura* and its use.
533. Describe the *eye*, regarded as an optical instrument.
534. What power of adjustment does the eye possess, and under what circumstances is this power called into action?
535. When is a person said to be *short-sighted*, and when *long-sighted?* What are the remedies for these defects, respectively?
536. What is the principle of the *simple microscope?* What condition must be answered in order that the virtual image which is formed may be distinct?
537. What are the essential parts of a *telescope?* Explain by a figure how a telescope forms a virtual and magnified image of a distant object.
538. What is the optical difference between the simple microscope and the telescope?

Lesson XXXI. — *Dispersion of Light by the Prism.*

539. What great discovery did Newton make as to the nature of white light?
540. Explain the dispersion of light by a prism. Give the seven principal colors of the spectrum in the order of refrangibility.
541. Why is it of great importance, in experiments upon the decomposition of light by prisms, to make use of a very narrow slit?
542. What is the method, employed in the *spectroscope*, of multiplying the dispersions of rays of different refrangibilities? Describe briefly the spectroscope of Gassiot.
543. Explain an optical method of recombining the various constituents of white light.
544. Explain a mechanical method of combining the various colors of the spectrum so as to form white light.

Lesson XXXII. — *Thermo-Pile.*

545. In what way can we compare together the intensity of a ray of light and a ray of dark heat, and obtain a true measure of the energy of these rays, provided instruments of sufficient delicacy be employed?

546. Explain the principle of the *thermo-pile* discovered by Seebeck.

547. In order to obtain by the use of thermo-electricity a very delicate instrument wherewith to measure radiant heat, what three objects must be accomplished?

548. How may a strong thermo-electric current be produced? Illustrate with a figure.

549. Describe *Thomson's galvanometer* and its action, explaining in particular how the magnetic force is overcome, and how by optical arrangements any small motion of the needle is very much magnified.

550. Explain the construction of the *thermo-pile.*

551. Suppose that when a source of radiant heat is placed before the pile, the luminous slit is made to move on the screen through twenty divisions of the scale; suppose, again, that when a different source of heat is presented to the pile, the index moves over forty divisions; what is the relation between the heating effects of the two sources? What is the general law?

552. Explain how, by the joint aid of the spectroscope and the pile, we are enabled to analyze a beam of sunlight so as to estimate the heating effects of all the different rays in the solar spectrum. What weak point is there in this method?

553. What source of heat did Leslie employ in his experiments on dark heat, and what was one of the results at which he arrived?

554. What are some of the facts established by Melloni with the aid of the pile? What is *diathermancy?*

555. Explain, by the aid of a figure, the manner in which Melloni performed his experiment to prove that dark heat is capable of refraction.

556. What two facts in reference to dark heat were established by Forbes ?

557. Give a sketch of the manner in which we may explore experimentally the heat spectrum produced by a heated strip of coal, for example.

558. Suppose we begin by heating a strip of carbon to a heat below redness, and producing a spectrum of the radiation from the carbon by means of a rock-salt prism ; trace the changes in this spectrum as the temperature of the carbon is gradually raised to a very high point.

559. What is the position upon the spectrum of *actinic* rays as they are called, and what power do they possess ?

560. Give a graphical representation of the sun's visible spectrum, locating the primary colors, and drawing the curve of intensity of light.

561. Give a graphical representation of the *entire* solar spectrum, and draw the curve of intensity of heat.

562. Where is the maximum luminous effect in the solar spectrum, and where the maximum heating effect ?

563. Explain the statement " *the spectrum of carbon is a continuous one.*" In general, what bodies give continuous spectra ?

564. In what respect do the spectra of gases differ from those of solid bodies ? What is the spectrum of ignited sodium vapor ? of thallium ?

565. What chromatic phenomena will be observed if we ignite a piece of metallic sodium in a dark room ?

566. In what way has electricity been found serviceable in spectrum-analysis ?

Lesson XXXIII. — *Radiation and Absorption.*

567. What great and striking difference between the spectra of solids and those of incandescent gases was made known in the last Lesson ?

568. Explain how it may be shown by experiment that at comparatively low temperatures, say 100° C., a lamp-black surface, or one of glass or white paper, radiates much more than a surface of polished silver.

569. Explain how the relative absorbing powers at 100° C. of the substances mentioned in the preceding question may be determined experimentally.

570. On comparing two tables, one containing the radiating and the other the absorbing powers of a series of substances, what general law comes to view ?

571. In what important respect do surfaces differ as regards their absorbing powers for different rays ? What instances of this can you give ?

572. Describe three experiments illustrative of the radiation from bodies of high temperature. What general relation between absorption and radiation do these experiments tend to establish ?

573. What is found to be the behavior of transparent colorless glass as regards absorption and radiation ; also, of a film or stratum of air ?

574. Give a generalization of the conclusions to be drawn from the preceding experiments.

575. Explain what is meant by *selective* or *partial* absorption by describing the behavior of white paper, and also of the glass bulb of a thermometer, at different temperatures.

576. What is it that makes the leaves of plants appear green ? In general, what is the physical cause of color ?

577. What familiar illustration of selective absorption is afforded by colored glasses ?

578. Describe an experiment in proof of the law that *bodies when cold absorb the same kind of rays that they give out when hot.*

579. Describe another experiment in proof of the above law.

580. Describe a third experiment in proof of the same law.

581. Suppose that we introduce into a chamber, kept uniformly at a white heat, transparent glass, polished platinum, coal, and black and white porcelain ; and that,

LESS. XXXIII.] ELEMENTARY PHYSICS. 51

after leaving them until they have acquired the temperature of the walls of the chamber, as a first experiment we simply examine them through a small hole; finally, suppose that, as a second experiment, we hastily withdraw the substances, and, without allowing them time to cool, examine them in the dark;—what will be the appearances presented in the two experiments, and how may they be reconciled with one another?

582. If we introduce red and green glass into a white-hot chamber, and then view them through a small opening, they will appear to have entirely lost their color. Explain this.

583. Give the grounds upon which black bodies have been selected as the standard or typical radiators.

584. What simple method may be employed to ascertain whether or not one body is hotter or colder than another?

585. Explain how, by the aid of the spectroscope, we may learn the chemical nature of a substance.

586. Show how spectrum analysis has demonstrated that there are present in the sun, in the state of vapor, various substances well known on the earth, as sodium, iron, zinc, magnesium, etc.

587. What conclusion may we draw from the results of Professor Tyndall's investigations into the absorption of various gases for dark heat?

588. Explain the part which the aqueous vapor of the atmosphere plays in relation to the heating effect of the sun upon the earth's surface.

589. Show how the laws of radiation explain the deposition of dew.

590. Give some examples of the phenomenon called *phosphorescence*. Also give an instance of the similar phenomenon known as *fluorescence*.

591. What is Professor Stokes's explanation of the phenomena of phosphorescence and of fluorescence? What is really the only difference between the two phenomena?

Lesson XXXIV. — *On the Nature of Radiant Energy.*

592. What two hypotheses regarding the nature of light were propounded by Newton and Huyghens, respectively?

593. What crucial test between these two hypotheses has been found?

594. What striking analogy is there between light and sound which leads to the belief that light must be a motion similar to sound, that is to say, undulatory?

595. On the undulatory theory, how does the eye distinguish between rays of different wave-lengths? What analogy is there in this particular between light and sound?

596. Illustrate what is meant by the *front* of a wave, and give a general definition of the same.

597. In what direction, relatively to its front, does a wave always proceed?

598. Deduce the law of reflection from the undulatory theory of light.

599. Deduce the law of refraction from the undulatory theory of light.

600. In the undulatory theory, what does the index of refraction of a substance represent?

601. What reason may be assigned why the velocity of light should be less in glass, for example, than in vacuo?

602. What analogy serves to aid the mind in perceiving why reflection and refraction accompany each other when light falls on a polished glass surface, for example?

603. What well-known facts respecting shadows might lead us to imagine that light differs from sound in a fundamental respect? What is the cause of the difference between sound-shadows and light-shadows?

604. Give instances of the manner in which the beautifully colored appearances due to the *interference* of light may be produced. What is the general explanation of these appearances according to the undulatory theory?

605. Explain *Newton's rings.*

606. Explain the colors of thin plates, such as those of a soap-bubble.

607. State an apparent objection to the undulatory theory, derived from the laws of energy, and show how this objection is entirely removed.

608. Explain why it is, that, when a sounding body is approaching the ear, its note is rendered more acute, while if it be receding from the ear, its note becomes more grave.

609. Show how Mr. Huggins has been able to make out the proper motions of several stars in a direction to and from the eye.

Lesson XXXV. — *Polarization of Light. Connection between Radiant Energy and the other Forms of Energy.*

610. What two kinds of wave-motion are met with in nature?

611. Which kind of vibrations is capable of assuming a particular side or direction, and how may this fact be illustrated?

612. What is the meaning of the term *polarization?*

613. Give an illustration to show how a mixture of vertical and horizontal waves may be sifted, so to speak, and deprived of the vertical components of the waves, or of the horizontal components, or of both.

614. Describe the action of tourmaline upon light.

615. What is the only possible explanation of the phenomena which are observed? To whom are we indebted for this explanation?

616. How is polarization by reflection effected, and what is meant by saying that the light is then "polarized in the plane of reflection"?

617. Show that an ordinary ray of light may be made to disappear entirely by two reflections.

618. Explain the *double refraction* of light by a crystal of Iceland spar. What is the appearance of a small body as seen through a piece of Iceland spar?

619. Explain the general connection between radiant energy, mechanical energy, and the energy of absorbed heat.

CHAPTER VII.

ELECTRICAL SEPARATION.

Lesson XXXVI. — *Development of Electricity.*

620. Mention two leading facts in the early history of electricity. From what is the word derived?

621. What marked difference exists between metal and glass as regards their power to conduct electricity? By what terms do we express this difference?

622. Give the tables of the most important conductors and insulators. What is the character of the transition from the one class of bodies to the other?

623. Why is it very desirable to make all experiments on electricity in a dry atmosphere?

624. Show by an experiment that there are two kinds of electricity, and give their names. When do electrified bodies attract each other, and when do they repel each other?

625. Explain the hypothesis of two fluids.

626. Give a table of twelve common substances in the order of their relative capacity for positive electrification.

627. Mention other modes of developing electrical separation besides friction.

628. What appears to be an essential condition for the production of electricity by the mutual action of two bodies?

629. What general connection is there between electrical separation and energy or mechanical work?

630. Describe the electrical properties of tourmaline. What species of energy is spent in this case to produce the electrical separation?

LESSON XXXVII. — *Measurement of Electricity.*

631. Show how an electrical charge upon a metallic body can be subdivided.

632. Describe Coulomb's *torsion-balance*, and the experiments which demonstrate the law of electrical action between two bodies, so far as it depends on the distance of the bodies from each other.

633. Explain how, by means of Coulomb's torsion-balance, we may prove the law of action between two electrified bodies, so far as it depends on the quantities of electricity upon the bodies.

634. What is a convenient unit of electrical force? Find in terms of this unit the force exercised by 6 units of positive upon 4 units of negative electricity at the distance 3.

635. Show, by an experiment, that electricity manifests itself only on the surface of bodies, and give the explanation of this fact.

636. In certain countries electrical manifestations are often produced by combing the hair, rubbing a silk dress, etc., while they are not observed in other parts of the world. How do you account for this?

637. In what way does a charge of electricity distribute itself on a sphere? on a pointed conductor? What accounts for the distribution in each case?

638. Define the term *electric density*. Describe a body such that the electric density will be much greater at some parts than at others.

639. Show how the relative distribution of electricity over the surface of a body may be ascertained by means of the *proof-plane*.

LESSON XXXVIII. — *Electrical Induction.*

640. What will happen if we bring near together two insulated conductors, one charged with electricity, and the other not charged? What is this kind of action called?

641. Suppose that the neutral conductor in the preceding question be divided into two parts, what will be the electrical condition of each part? How may this fact be proved by experiment?

642. Suppose that we slowly bring a conductor, charged with electricity, towards another conductor not charged, until they are very near each other; explain the phenomena which will take place.

643. How may it be rendered evident that the inductive effect of electricity depends on the distance between the two conductors?

644. What new light does electrical induction throw upon the fact that electricity only shows itself at the surfaces of bodies?

645. What important fact was discovered by Faraday in his researches upon electrical induction? What is the *inductive capacity* of a substance?

LESSON XXXIX. — *Electrical Machines, etc.*

646. Of what two parts is every electrical machine composed?

647. Describe the *plate electrical machine*, and explain its action.

648. Describe the simple experiments with an electrical machine which may be performed, —

    1. by holding the finger near the charged conductor;

    2. by placing an individual on an insulating stool;

and give the explanation of the experiments according to the two-fluid theory.

649. Describe the *electrophorus*, and explain its action.

650. Describe the *gold-leaf electroscope*, and explain how it enables us not only to detect the presence of electricity, but also to determine whether the electricity is positive or negative.

651. What difference between an *electroscope* and an *electrometer* is indicated by the derivation of the two words themselves?

652. Explain the method of measuring electrical charges employed by Sir W. Thomson in his electrometers.

653. Explain the accumulation of electricity by *condensers*. If the condensing plates are separated, the pithballs attached to them will diverge; explain this.

654. Describe the *Leyden jar*. Show how it may be charged and discharged, and explain its mode of action.

655. If a Leyden jar be allowed to stand for a short time after being discharged, it is found that it has a small residual charge left in it; what is the probable explanation of this?

656. What is an *electric battery*, and how formed from its component parts?

657. What knowledge of the nature of the electric spark has been obtained by viewing it through the spectroscope, and what use has been made of this knowledge?

658. What transmutation of energy do we have in the electric spark?

659. Investigate the relation between the charge of a Leyden jar and the amount of heat produced by discharging the jar, showing that the whole heating effect will be proportional to the square of the quantity of electricity divided by the surface of the jar.

660. How can it be shown, experimentally, that the duration of the electric spark is exceedingly short?

661. How has Sir C. Wheatstone succeeded in measuring the duration of the electric spark? What was the result of his experiments? What did he also find to be the velocity of electricity?

662. Who first proved that lightning is only a manifestation of electricity on a large scale, and in what way did he prove that this is the case? What advantage has been taken of this knowledge?

663. To what are the following phenomena due?—
1. The light which constitutes the electric flash.
2. The noise which accompanies the same.
3. Its destructive effect in rending substances.

664. If we bring a hollow insulated brass ball near an electric machine in action, we shall get a spark, but it will be very feeble. If, however, we touch with our finger that part of the conductor which is farthest from the machine, or make a connection between this conductor and the ground, the spark from the machine will be much more intense. Explain this.

665. Show how to obtain from an insulated conductor, near an electric machine in action, —
  1. a series of sparks or shocks ;
  2. a continuous rush of electricity.

666. Explain the efficacy of lightning-conductors.

667. Discuss briefly the connection between electrical separation and the other forms of energy.

CHAPTER VIII.

ELECTRICITY IN MOTION.

LESSON XL. — *Magnetism.*

668. What is the origin of the term *Magnet?*

669. Describe some of the properties of a magnet. What are its poles, and how are they distinguished from each other?

670. What is the difference between magnetic and diamagnetic bodies? Enumerate the most important bodies of each class. In what respect does iron stand alone?

671. Explain the behavior of magnetic and diamagnetic bodies when suspended midway between the two poles of a powerful magnet.

672. Explain the behavior of magnetic and diamagnetic bodies when suspended between the poles of a magnet in a magnetic liquid instead of in air.

673. What is the law of the mutual action of magnetic poles?

674. State the quantitative law of force in magnetic attractions and repulsions. By whom was this law discovered?

675. Prove that, in consequence of this law, if we suspend a small magnet by a thread and cause it to approach the pole of a powerful magnet, the small magnet will exhibit a tendency to rush bodily to the large magnet; and find the measure of this tendency.

676. Explain magnetic induction.

677. Describe the effect of breaking a magnet, and give a theory of the distribution of the magnetic fluids which will explain the properties of magnets, both when entire and when broken.

678. What difference is there between soft iron and

hard steel as regards susceptibility to magnetism? Explain one mode of magnetizing a steel bar. What is the effect of heat on magnets?

679. If we were to suspend a magnetic needle in such a manner that it was perfectly free to move in any direction, how would it place itself?

680. What are *magnetic meridians?*

681. What facts are stated as showing that a magnetic needle will not everywhere and always point as it does in Great Britain at the present moment?

682. At what places on the earth's surface is a magnetic needle of no use to the mariner, and why?

683. Explain why the effect of the earth's magnetism upon a magnetic needle is merely directive.

Lesson XLI. — *Voltaic Batteries.*

684. What was the famous phenomenon first observed by Galvani in 1786, and how was it explained by Galvani and by Volta respectively?

685. Explain the construction and mode of action of *Volta's pile.*

686. Describe the arrangement known as *Volta's crown of cups.*

687. How did Volta explain the effect produced by the *voltaic battery?*

688. Illustrate the manner in which the total effect produced by Volta's pile depends on the number of elements in the pile.

689. Explain in what way the contact theory as held by Volta is inconsistent with the laws of energy.

690. What is the chemical theory of the action of the voltaic battery?

691. What was the nature of the crucial experiment made by Sir W. Thomson in reference to the two theories of the voltaic battery, and what conclusions are to be drawn from it?

692. What is denoted by the term *electromotive force*?

693. What results did Sir C. Wheatstone obtain in his experiments on the electromotive force in different combinations of platinum, zinc, and potassium, and what general law do they illustrate?

694. Classify the metals according to their order in the electromotive series.

695. What two causes greatly enfeeble a single-liquid battery after it has been in action a short time?

696. Describe Daniell's constant battery, and its mode of action.

697. What are the advantages of amalgamating the zinc plates?

698. Describe Grove's battery, and its mode of action.

699. How may the existence of a *thermo-electric current* be easily demonstrated? Why are the metals bismuth and antimony generally used in thermo-electric combinations?

700. Illustrate the application of the law of Art. 374 in thermo-electric combinations.

701. Within certain limits what is the strength of a thermo-electric current proportional to? But what has Cumming shown in the case of copper and iron?

Lesson XLII.—*Effect of the Electric Current upon a Magnet.*

702. When, and by whom, was the important discovery of the connection between an electric current and a magnet made?

703. Explain the nature of Oersted's experiment.

704. State the rule which expresses the relation between the behavior of the needle and the position and direction of the current, and apply this rule to the four distinct cases which are possible.

705. What is the object of a *galvanometer*? Explain the construction and mode of action of a single-needle galvanometer.

706. Explain the construction and mode of action of an *astatic* galvanometer. Describe the mirror arrangement for increasing the sensibility of a galvanometer.

707. Upon what law does the action of a current on a needle depend?

708. Describe the *tangent compass* and its action. Why has the instrument received this name?

709. Describe an *electro-magnet*. How do they compare in strength with natural magnets?

710. What curious facts have been observed in the magnetization of soft iron bars?

711. State the principle of the electric telegraph. What takes the place of a return wire in electric telegraphs, and what advantages are gained by this substitution?

Lesson XLIII. — *Action of Currents on One Another, and Action of Magnets on Currents.*

712. What are the chief laws of the mutual action of electrical currents?

713. Discuss the various cases which may arise under Law III.

714. Explain a case in which a continuous rotation of currents is produced by their mutual action.

715. Examine the action of the earth's magnetism upon a circular vertical current which is free to place itself in any position.

716. Explain the construction and behavior of a solenoid.

717. State Ampère's hypothesis concerning magnetism, and show that it explains the known relations between magnets and currents, and between magnets and magnets.

718. What instance of the conservation of energy do we have in the case of two similar voltaic batteries, each charged with the same amount of zinc, if one battery is made to do external work, while the other does no external work at all?

LESSON XLIV. — *Induction of Currents.*

719. State the laws of the induction of electric currents. Who discovered current induction?

720. Explain how magnets may be made to play the part of currents in the phenomena of induction.

721. Show that the phenomena of induction are in harmony with the laws of energy.

722. Describe a method, employed by Joule, for converting mechanical energy into that of induced currents, and from that into heat.

723. What two kinds of electrical machines are there which depend for their action on the laws of induction?

724. What is the principle of a *magneto-electrical* machine? What arrangement is employed in Clark's machine? What is the object of a *commutator*? For what purposes, among others, are these machines used?

725. Describe *Ruhmkorff's coil,* and explain its mode of action.

LESSON XLV. — *Distribution and Movement of Electricity in a Voltaic Battery.*

726. Who first developed the laws regulating the motion and distribution of electricity in a battery?

727. State the laws which regulate the electro-motive force of a battery.

728. Investigate the subject of electrical resistance in a manner similar to that employed in studying thermal conductivity, and deduce Ohm's formula for expressing the relation between the intensity of the current, the electromotive force, and the resistance of a galvanic circuit.

729. Upon what three things does the electrical resistance of a substance depend?

730. Modify the fundamental formula of Ohm so as to express the intensity of the current in a battery of ten cells with a definite external resistance.

LESS. XLVI.]   *ELEMENTARY PHYSICS.*   65

731. Examine, by means of Ohm's formula, the effect of increasing the number of cells in a battery ;
   1. when there is no external resistance ;
   2. when the external resistance is small compared with the internal;
   3. when the external resistance is large compared with the internal.
732. Why is it necessary to have a large number of cells in order to produce the electric light ?
733. Why is it advantageous to multiply the number of couples in a thermo-electric current ?
734. Explain, by the aid of Ohm's formula, the effect of increasing the size of the plates in a voltaic battery.
735. What arrangement in a battery is preferable, when the battery is to be used to produce thermal effects ? Why ?
736. What law, as to the intensity of the current in different portions of a circuit, is likewise due to Ohm ?
737. Explain a method of comparing the resistance (and hence the conductivity) of metallic wires by means of a galvanometer.
738. What points of resemblance have been observed between the electric and the thermal conductivities of substances ?

LESSON XLVI. — *Effects of the Electric Current.*

739. Compare, as regards quantity, tension and the resultant physiological effects, the Leyden jar battery, the voltaic battery, and a flash of lightning.
740. When an electric current is made to pass through a circuit, to what is the heating effect of the current proportional ?
741. Show that the increase of temperature produced by the passage of the same quantity of electricity through a wire will vary inversely as the square of the cross section of the wire.

E

742. Deduce from the above law, that the heat generated in a given time is proportional to the square of the intensity of the current.

743. If one part of a circuit be composed of a metre of silver wire two square millimetres in cross section, and another of five metres of zinc wire four square millimetres in cross section, show that the relative heating effects of the current on these two wires will be as $1 : 8{\cdot}62$.

Specific electric conductivity of silver, 100; of zinc, 29.

744. Looking at the subject from the stand-point of the doctrine of energy, in what consists the difference between dissolving zinc by acid in an ordinary vessel and doing so by the voltaic arrangement?

745. Describe how the *electric light* is produced.

746. Define the terms *electrolysis, electrolyte*.

747. Describe a voltaic arrangement which may be employed to decompose water.

748. What is the distinction between *electro-positive* and *electro-negative* elements?

749. In the electrolytic decomposition of water, for example, the question naturally arises, Is the oxygen of each molecule which is decomposed carried bodily to the one pole, and the hydrogen to the other? What was Davy's test experiment upon this point?

750. Explain Grotthuss's hypothesis.

751. State the laws of electrolytic action discovered by Faraday.

752. Explain the principle of the *electrotype process*.

753. What is the effect of passing polarized light through glass subjected to the action of a powerful electro-magnet?

754. Under what conditions are peculiar stratifications of light and colors produced by the current?

755. Will a current pass through a perfect vacuum?

756. What is the cause of the peculiar smell which is often noticed when an electric machine is in action?

CHAPTER IX.

ENERGY OF CHEMICAL SEPARATION.

Lesson XLVII. — *Concluding Remarks.*

757. Why is it natural to expect that a definite amount of carbon will, when burnt, always furnish a definite amount of heat?

758. Who have investigated the quantity of heat given out in chemical combination?

759. To what general result was Andrew led by studying the heat given out during the mutual action of metals?

760. What grounds are there for believing that the electro-motive forces are really those which cause heat when chemical combination takes place?

761. What relation have we found to exist between the doctrine of the conservation of energy and the chimera of perpetual motion?

762. In what way might a champion of perpetual motion assent to the doctrine of the conservation of energy without absolutely giving up his cause?

763. Give the outlines of the doctrine of the dissipation of energy, and show what bearing this doctrine has on the problem of perpetual motion.

764. Trace back the energy of our system through its various transmutations to its ultimate source.

765. What vast store of energy was provided by Nature in geological ages?

766. Show that water-power and wind-power are really products of the sun's rays.

767. What single small exception is there to the statement that "all the work done in the world is due to the sun"?

768. What would seem to be the answer which we must give to the question, Will the sun last forever?

769. In fine, to what ultimate conclusion does the principle of degradation conduct us?

770. Enumerate the various kinds of energy which have been studied in this book.

771. Recapitulate various instances of the transmutation of visible kinetic energy.

772. Give examples of the conversion of visible potential energy.

773. Enumerate the instances of the transmutation of heat.

774. Give instances of the transmutation of radiant energy.

775. Mention examples of the transformation of the energy of electrical separation.

776. Give examples of the conversion of the energy of electricity in motion.

777. Give instances of the transmutation of the energy of chemical separation.

PART II.
EXERCISES AND PROBLEMS.

THE Exercises in large type are, in the main, direct and simple applications of, or deductions from, the principles of the text-book : in the cases in which special difficulties might arise or in which new definitions are introduced, hints or explanations will be found in Part III. (Answers and Solutions). These Exercises demand only a fair knowledge of the Elements of Arithmetic, Algebra, and Plane Geometry.

The Exercises printed in smaller type are intended to be more difficult than the others, and some of them involve principles which are not explicitly stated in the text-book. They are designed chiefly for use with advanced sections or as voluntary exercises. Those who wish to take them should have an elementary knowledge of Plane Trigonometry and of Analytic Geometry. In a few cases a knowledge of the Calculus may perhaps be serviceable, although it is not required. A summary of mathematical data and formulae is given in Appendix IV. When aid is required, it must be obtained from competent teachers, or from books. Appended is a list of elementary works which may be consulted with advantage, particularly the first two and the last two.

THOMSON AND TAIT'S *Elements of Natural Philosophy*, Part I. (London & New York: McMillan & Co.)
KERR's *Rational Mechanics.* (Glasgow : W. Hamilton.)
GOODWIN's *Elementary Statics, and Elementary Dynamics.* (Cambridge, England : Deighton, Bell, & Co.)
BESANT's *Elementary Hydrostatics.* (Cambridge, England : Deighton, Bell, & Co.)
HAUGHTON's *Manual of Mechanics.* (London & New York : Cassell, Petter, & Galpin.)
TODHUNTER's *Mechanics For Beginners.* (London & New York : McMillan & Co.)
GOODEVE's *Principles of Mechanics.* (London : Longmans, Green, & Co.)
BURAT, *Précis de Mécanique.* (Paris : Victor Masson et Fils.)
BRIOT, *Leçons de Mécanique.* (Paris : Dunod, Éditeur.)
BRESSE ET ANDRÉ, *Cours de Physique*, les 2 premier fascicules. (Paris : Dunod, Éditeur.)
DESCHANNEL's *Natural Philosophy*, Translated by EVERETT, Part I. (New York ; D. Appleton & Co.)

INTRODUCTION.

1. Give an illustration, not mentioned by the author, of relative motion.

2. Give an illustration, not mentioned by the author, of force producing motion ; also, of force stopping motion.

3. Give an example of forces in equilibrium.

4. If you are running towards the North, and, as suddenly as possible, change the direction of your motion from the North to the East, do you think that force is expended in producing this change ?

5. Can you give an instance of a body which is not acted upon by any force whatever ?

6. Mention an object which is known to us through the medium of a single sense ; also, an object which is known through the medium of more than one sense.

7. Explain and illustrate the distinction between a *phenomenon* and a *law of Nature*.

8. What distinction can you draw between a *body* and a *substance*.

9. Two steamers are moving with equal velocities in the same direction. A passenger on one steamer looks at the other from his state-room window ; how will it appear to him ? Suppose the other steamer suddenly appears to change its velocity ; in what two ways might this phenomenon be produced ?

CHAPTER I.

LAWS OF MOTION.

LESSON I. — *Determination of Units.*

10. How many square feet are there in 124 acres?

11. How many square decimetres are there in 124 ares?

12. Reduce 346768595 cubic inches to cubic yards.

13. Reduce 346768595 cubic centimetres to cubic metres.

14. How many litres are there in 2 steres?

15. Reduce 6,000,000 grammes to tonnes.

16. Reduce 218·75 grains to grammes.

17. What ratio exists between a cubic centimetre and a cubic metre?

18. What is the weight of 64 litres of water? of 64 cubic centimetres of water?

19. Show that the number which expresses the volume in litres of a quantity of water also denotes the mass in kilogrammes. Examine also the case in which the volume of the water is expressed in cubic centimetres.

20. A rectangular trough is 12 metres long, 2 metres wide, and 80 centimetres deep. How many kilogrammes of water will it hold?

21. Define *density*. What is the numerical measure of the density of a substance?

22. Prove that the mass of a body is equal to the product of its volume and its density; or, if V denotes the volume, D the density, and M the mass, that $M = VD$.

23. Why is the density of water equal to unity in the Metric System?

24. Find the mass of 74 litres of cork (density of cork, 0.24).

72   ELEMENTARY PHYSICS.   [CHAP. I.

25. Prove that the densities of two bodies are proportional to the masses of equal volumes of the bodies.

26. Explain the distinction between *mass* and *weight*.

27. A ship sails 504 miles in a week. Find the *average* velocity in miles per hour.

28. Compare the velocities of two points which move uniformly, one through 5 feet in half a second, the other through 100 yards in a minute.

29. The daily rotation of the earth is uniform. Taking its circumference as 25,000 miles, determine the velocity of a point on the equator.

30. A body is moving with a velocity of 30 feet per second. With what velocity must another body move, which starts from a given point 3 minutes after the former and overtakes it in 10 minutes?

31. For 6 seconds a body moves with a velocity of 10, and for the next 9 seconds with a velocity of 15. What uniform velocity would have carried it over the same space in the same time?

32. Compare the velocities of two points, one of which moves uniformly around the circumference of a circle in the same time that the other moves along the diameter.

33. The height of a cylindrical cistern is 12 metres and its diameter is 6·5 metres. How many kilogrammes of water will it hold?

34. One litre of a substance weighs 280 grammes, and a piece of another substance twice as dense as the first weighs 400 grammes. Find the volume of the second substance.

35. Find the ratio of the kilometre to the nautical mile or *knot*.

36. Show that the proper measure of density is the mass of unit of volume.

37. If the unit of mass be increased a times, and the unit of volume be increased b times, how will the measure of density be altered?

38. A cylindrical log of wood, a metres long, and b centimetres in diameter, weighs c kilogrammes. Compare its density with that of a substance the density of which is known to be d.

39. What are the *dimensions* of velocity in terms of the fundamental units of length and time.

40. Define angular velocity. What is the unit of angular velocity, and what are its dimensions in terms of the fundamental units?

41. Find the linear velocity with which a point must move on the circumference of a circle in order to describe one unit of angular velocity per second.

42. What is the measure of the angular velocity of the hour hand of a clock? of the minute hand?

43. Prove that the linear and angular velocities of a point moving on the circumference of a circle are connected by the equation, $v = r\,\omega$, in which v denotes linear velocity, r radius of circle, and ω angular velocity.

44. Two bodies begin to move uniformly at the same time along the same line, the first from a point A with a velocity v, the second from a point B with a velocity v': —

(1) How far apart will they be at the end of t seconds?
(2) When will they be together?
(3) How far will they be from A when they are together?

45. If a velocity be expressed by 6 when one second is taken as the unit of time, what would be its measure if one minute were taken as the unit of time?

46. If v denotes a velocity in the metre-second system, prove that the same velocity will be denoted by $\dfrac{n\,v}{m}$, in a system in which the unit of length is m metres and the unit of time n seconds.

47. Prove that if two points move uniformly with any velocities in fixed directions, the line joining the points will always remain parallel to itself.

Lesson II. — *First Law of Motion.*

48. When we find a body moving uniformly and in one constant direction, what may we infer with regard to the total force that is acting upon the body?

49. Illustrate the principle of Inertia by reference to the condition in which a person finds himself when standing in a boat at starting or stopping.

50. Account for the practical rule which habit teaches us to observe in jumping from a carriage which is in motion.

51. Serious accidents have sometimes happened by carriages oversetting when moving along a sharp curve in

74   ELEMENTARY PHYSICS.   [CHAP. I.

the road. Explain the cause of these accidents, and show how they might have been prevented.

52. Show that the First Law of Motion contains the convention universally adopted for the measurement of *Time*.
53. What definition of *Force* does the First Law of Motion give us?
54. Review briefly the evidence in favor of the truth of the First Law of Motion.

LESSON III. — *Second Law of Motion. Motion produced by Gravity. Kinematics.*

[In the problems upon the motion produced by gravity the resistance of the air is neglected, and g is to be taken as equal to 32·2 feet, or 9·8 metres.]

55. Give an additional illustration of the action of a single force on a moving body.

56. State in general terms the rule for compounding two simultaneous motions or velocities in different directions, — a rule or proposition known as the *Parallelogram of Velocities*, — and give a demonstration of the same.

57. If a man is rowing a boat directly across a river two miles wide at the rate of four miles an hour, and the current at the same time is taking the boat down stream at the rate of three miles an hour, find, —

(1) In what direction the boat will move;
(2) How far it will have gone when it reaches the opposite bank;
(3) How far the landing-place will be from the point directly opposite the starting-place;
(4) How long the boat is in motion;
(5) How long it would have taken to cross the river if there had been no current.

58. Explain the geometrical method of finding a single velocity equivalent to any number of simultaneous velocities.

59. Explain the geometrical method of *resolving* a velocity in a given direction into two *component* velocities in any given directions.

60. Prove that if velocities represented by the sides of a triangle *taken in the same order* be impressed simultaneously upon a point it will remain at rest.

61. Prove that the resultant of velocities represented by the sides of any closed polygon whatever, taken all in the same order, is zero.

62. Show that the *value,* or *resolved part,* or *effective component* of a known velocity, estimated along a given line, is the *projection* of the line representing the velocity upon the given line. Examine the case in which the given line makes a right angle with the line representing the velocity.

63. Suppose that six forces act simultaneously upon a body, such that separately they would impart to it the following velocities: —

4 feet per second towards the East,
9 "   "   "   "   "   North,
2 "   "   "   "   "   "
7 "   "   "   "   "   West,
5 "   "   "   "   "   "
3 "   "   "   "   "   South.

Find the magnitude and direction of the resultant velocity.

64. If a cannon-ball were discharged from the rear end of an express-train, directly along the track, at the same rate as the train is moving forwards, what would be the motion of the ball relative to the ground?

65. Expose the fallacy in the following specimen of erroneous mechanical reasoning: —

"Let the ball be thrown *upwards* from the mast-head of a *stationary ship,* and it will fall back to the mast-head, and pass downwards to the foot of the mast. The same result would follow if the ball were thrown upwards from the mouth of a mine, or the top of a tower, on a *stationary earth.* Now put the ship *in motion*, and let the ball be thrown *upwards*. It will, as in the first instance, partake of the two motions, — the upward or vertical A C, and the horizontal A B, as shown in Fig. 47; but

FIG. 47.

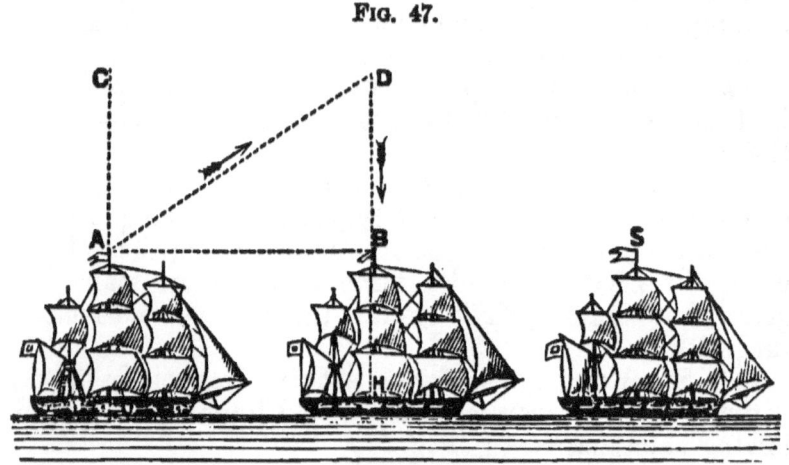

because the two motions act conjointly, the ball will take the diagonal direction A D. By the time the ball has arrived at D, the ship will have reached the position B; and now, as the two forces will have been expended, the ball will begin to fall, by the force of gravity alone, in the vertical direction D B H; but during its fall towards H, the ship will have passed on to the position S, leaving the ball at H, a given distance behind it." *

66. Bishop Wilkins, an English divine of the 17th century, and author of a Treatise on the Art of Flying, proposed the following "new and easy way of travelling." A large balloon was to be constructed and provided with apparatus to work against the varying currents of the air. The balloon, having been allowed to ascend to a convenient height, was to be kept practically at rest by working the apparatus, while the earth revolved beneath it ; and when the desired locality came in view, those in the bal-

* Earth Not a Globe, by "Parallax," pp. 64, 65. London: John B. Day. 1873.

loon were to let out gas and drop down at once to the earth's surface. In this way New York, for example, would be reached from London in a few hours, or rather New York would reach the balloon at the rate of more than 700 miles an hour.

Show the futility of any such method of travelling.

67. Define acceleration, uniform acceleration, variable acceleration. How is uniform acceleration measured ?

68. Find the *average* acceleration of a point the velocity of which increases from ten miles per hour to sixty miles per hour in two hours.

If at the end of the first hour the velocity is fifty miles per hour, find the average acceleration during each hour.

69. A constant force acting thirteen seconds produces a velocity of four miles per hour. Find the acceleration.

70. A body falls to the ground from rest in 6 seconds ; find the space passed over.

71. Through what space will a body fall in the ninth second of its descent ?

72. A stone strikes the ground with a velocity of 98 metres per second. Find the height fallen through.

73. How long must a body fall to acquire a velocity of 322 feet per second ?

74. A rocket begins to ascend vertically with a velocity of 161 feet per second. How high will it rise, and what time will be occupied in both ascent and descent ?

75. Deduce and explain the general formula for determining the final velocity v of a body which, having an initial velocity a, is acted on by gravity for t seconds, viz. :—
$$v = a + gt.$$

76. A body is thrown downward with a velocity of 160 feet per second ; find its velocity at the end of five seconds.

77. A body is thrown upward with a velocity of 49 metres per second ; with what velocity and in what direction will it be moving at the end of 7 seconds ?

78. If a body have an initial velocity a, prove that the space passed over in t seconds under the action of gravity is given by the equation, —

$$s = at + \tfrac{1}{2}gt^2.$$

79. From the formulæ of Exercises **75** and **78** deduce the following, —

$$v^2 = 2gs + a^2$$
$$s = \tfrac{1}{2}(v+a)t$$

80. Prove that the time of ascent of a body projected vertically upward with the velocity a is $\dfrac{a}{g}$, and that the height ascended is $\dfrac{a^2}{2g}$.

81. Prove that the times of ascent and of descent of a body thrown upward are the same.

82. Prove that if a body is projected vertically upward it returns to the ground with the same velocity as that with which it was projected.

83. A cannon-ball, fired vertically upward, returned to the ground in 20 seconds; find the height ascended and the velocity of projection.

84. With what velocity must a stone be thrown down a well 100 metres deep, in order that it may reach the bottom of the well in one second?

85. Analyze geometrically the action of the wind on the sails of a vessel, and explain how it is possible for a vessel to sail nearly against the wind.

86. Explain the action of the wind on a kite.

87. A man can row a boat at a certain rate, and the current of a river is flowing at a certain rate; find the direction in which the boat should be steered in order that it may be rowed directly across the river.

88. Solve the problem of compounding two velocities by the trigonometrical method. Let u and v denote the velocities, w their resultant, ϕ, α, β, the angles between the direction of u and v, u and w, v and w, respectively; then the problem is, — given u, v, and ϕ, prove that, —

$$w^2 = u^2 + v^2 + 2 u v \cos \phi.$$
$$\sin a = \frac{v}{w} \sin \phi.$$
$$\sin \beta = \frac{u}{w} \sin \phi.$$

89. Deduce formulæ from those in the preceding problem for *resolving* a given velocity (w) into two velocities (u, v), making any assigned angles (a, β) with the given velocity.

90. The value of two velocities are 36 and 60, and the angle between their directions is 54°; find the resultant velocity, and the angles which its direction makes with those of the given velocities.

91. Two equal velocities are simultaneously impressed upon a body, one towards the north, the other towards the east. The resultant velocity is equal to 10. Find the two velocities, and the direction in which the body will move.

92. Explain, on kinematical principles, why the northern trade-wind appears to blow from a northeasterly direction, and the southern trade-wind from a southeasterly direction.

93. A person travelling due east at the rate of 4 miles an hour observes that the wind seems to blow directly from the south; and that, on doubling his speed, it appears to blow from the southeast. Find the velocity and the direction of the wind.

94. A deer is running at the rate of 20 miles an hour, and a sportsman fires at him when he is at the nearest point, 200 yards distant; what allowance should be made in taking aim, supposing the velocity of the rifle bullet to be 1000 feet per second?

95. A particle descends vertically *along the axis* of a tube which at the same time is carried forward in the horizontal direction, both motions being uniform; find the inclination of the tube from the vertical line.

96. Show that, in the case of variable velocity, the equation $v = \frac{s}{t}$ (which expresses the definition of the average velocity for any time) is more and more nearly true as the interval of time is taken smaller and smaller. Thence obtain the true measure of variable velocity.

97. Prove that the dimensions of acceleration are $\frac{L}{T^2}$.

98. If g denote an acceleration when the second is the unit of time and the foot is the unit of length, then, if we take m seconds as the unit of time and n feet as the unit of length, the *same* acceleration will be denoted by $\frac{m^2}{n} g$.

80   ELEMENTARY PHYSICS.   [CHAP. I.

99. Find the measure of the acceleration of gravity when one minute is taken as the unit of time.

100. A stone is thrown vertically upward with a velocity $3g$; find at what time its height will be $4g$, and its velocity at this time.

101. A stone is dropped into a well, and is heard to strike the surface of the water after 4·5 seconds; find the distance to the surface of the water, knowing that the velocity of sound is 340 metres per second.

102. A body is dropped from a height of 100 feet, and at the same moment another body is projected vertically from the ground: they meet half-way. What was the velocity of projection of the second body?

103. A body is projected vertically with a velocity of 30 metres per second. A second later another body is projected vertically from the same point with a velocity of 40 metres per second. When and where will the two meet?

104. With what velocity must a body be projected downwards that in n seconds it may overtake another body which has already fallen from the same point through a distance of a feet?

105. Prove that, when we take into account the resistance of the air, the time of ascent of a body projected vertically upward is *less* than the time of descent; and that the velocity on reaching the ground is *less* than the velocity on starting.

106. Prove that the velocity acquired in sliding down a smooth inclined plane is the same that would be acquired in falling freely through the vertical height of the plane.

107. Prove that the time of falling from rest down any chord of a vertical circle, drawn either from the highest or the lowest point of the circle, is constant.

108. Find the straight line of quickest descent from a given point to a given straight line.

109. If a be the base of an inclined plane, find the height in order that the time of descent may be a minimum.

PROJECTILES.

110. Prove that a body projected in any direction not vertical, and acted on by gravity, will describe a parabola.

111. Find the velocity at any point of the path of a projectile.

112. Determine the position of the focus of the parabola described by a projectile.

113. Determine the *latus rectum* of the parabola described by a projectile.

114. Find the maximum height reached by a projectile.

115. Find the whole time of flight of a projectile, and show that the times of ascent and of descent are equal.

116. Determine the range of a projectile on a horizontal plane through the point of projection.

117. Prove that the maximum range along a horizontal plane for a given velocity of projection is obtained by making the angle of projection equal to 45°, and is equal to the square of the velocity divided by g.

118. Find the horizontal range of a shell fired at an angle of 45° with a velocity of 500 feet per second.

119. Two bodies are projected simultaneously from the same point with different velocities and in different directions; find their distance apart at the end of a given time.

120. Find the velocity and direction of projection in order that a projectile may pass horizontally through a given point.

121. Find the velocity with which a body must be projected in a given direction from the top of a tower so as to strike the ground at a given point.

122. A shell is to be fired from the top of a cliff 300 feet high with a velocity of 600 feet per second, so as to strike a ship at anchor 600 yards from the base of the cliff. What must be the elevation of the gun?

123. Two bodies are projected from the top of a tower at the same instant, the one vertically upward with a velocity of 100 feet per second, and the other horizontally with a velocity of 60 feet per second; find their distance apart at the end of 2 seconds.

CURVILINEAR MOTION.

124. Taking acceleration in its expanded sense, .viz., *rate of change of velocity whether the change takes place in the direction of motion or not*, illustrate what is meant by change of velocity, and show how a curve may be drawn which shall represent the direction of the acceleration of a moving point at every instant.

125. If a point describe a circle, of radius r, with the uniform velocity v, prove that the acceleration is directed towards the centre, and is equal to $\dfrac{v^2}{r}$.

126. A stone is whirled round at the end of a string 2 metres long with a velocity of 12 metres per second; find the acceleration.

127. If T denote the time of revolution or *period* of a point which describes a circle with uniform velocity v, prove that —

$$\frac{\text{acceleration}}{\text{radius}} = \frac{4\pi^2}{T^2}.$$

128. If ω denote *angular velocity* in uniform circular motion, and T the period, prove that

$$\omega^2 = \frac{4\pi^2}{T^2} = \frac{\text{acceleration}}{\text{radius}}.$$

129. Distinguish between *tangential* and *normal* acceleration. Examine and illustrate the cases, (1) in which there is tangential but no normal acceleration, (2) in which there is normal but no tangential acceleration, (3) in which there is neither tangential nor normal acceleration.

SIMPLE HARMONIC MOTION.

130. Define simple harmonic motion. Give illustrations.

131. Define the following terms relating to simple harmonic motion, — *amplitude, displacement, period, phase, epoch.*

132. Prove that the velocity of a point executing a simple harmonic motion is at any instant directly proportional to the displacement at a quarter of a period earlier in phase.
When is the velocity a maximum? When a minimum?

133. Prove that the acceleration of a point executing a simple harmonic motion is at any instant directed towards the middle point, and is directly proportional to the displacement.
When is the acceleration a maximum? When a minimum?

134. If T denote the period of a simple harmonic motion, and ω the angular velocity on the circle of reference, prove that at any time, —

$$\frac{\text{acceleration}}{\text{displacement}} = \frac{4\pi^2}{T^2} = \omega^2.$$

LESSON IV. — *Second Law of Motion. Measure of Force.*

135. State the Second Law of Motion in the exact language employed by Newton.

136. Explain the meaning of the words "*change of motion*" in Newton's Second Law.

137. State, in the most general form, the Principle of the Independence of the Effects of Forces, which is contained in the second clause of the Second Law of Motion as given by Newton. Give examples which tend to establish the truth of this Principle.

LESS. IV.] *EXERCISES AND PROBLEMS.* 83

138. Why does it follow from the Second Law that the proper dynamical measure of a force is *the product of mass and acceleration*, or *the momentum generated in unit of time*.

139. Define the dynamical unit of force, first in general terms, and then in terms which express the British and the Metric values respectively.

140. Find the ratio between the British and Metric dynamical units of force.

141. Prove that one dynamical unit of force is equal to one statical unit of force (the pound or kilogramme) divided by the value of g at the given locality. Hence derive rules for reducing statical forces or pressures to dynamical measure, and *vice versa*.

142. Distinguish between the two uses of the word *pound* (or *kilogramme*).

143. Explain the employment of spring balances and common balances in estimating masses and forces.

144. The dynamical measure of a force (in Metric units) is 7252; reduce this to Paris kilogrammes of force (at Paris, $g = 9\cdot 81$ metres).

145. A mass of 80 kilogrammes at Paris is moved by a constant force which generates in one second a velocity of 6 metres per second. Find the measure of the force, (1) in dynamical units, (2) in kilogrammes.

146. On a mass of 140 lbs. at New York (where $g = 32\cdot 16$ feet) a force of 10 lbs. is constantly acting; find the acceleration.

147. The mass of an iron ball in pounds is m. Find its mass in kilogrammes; its weight at London in British statical units; its weight at Paris in Metric statical units; its weight in British dynamical units at London, where $g = 32\cdot 19$ feet; its weight in Metric dynamical units at Paris, where $g = 9\cdot 81$ metres.

148. From what height must the ram of a pile-driver, weighing 16 cwt., descend upon the head of a pile, in order that it may strike it with a momentum equal to that of a 42 lb. shot, fired with a velocity of 1610 feet per second?

84   ELEMENTARY PHYSICS.   [CHAP. I.

149. Given the mass and acceleration of a body, find the rate at which the momentum of the body is changing.

150. A pressure of 2 tons acts upon a railway train for 10 minutes; find the momentum acquired by the train.

151. How long must a force of 100 kilogrammes act on a mass of 2000 kilogrammes to impress upon it a velocity of 5 metres per second? What must be the magnitude of the force which would bring the body to rest in one second?

152. Prove that two forces are to each other as the accelerations which they impress upon the same or upon equal masses.

153. Show that the following proposition follows immediately from the Second Law: *The resultant of any number of forces applied at one point is found by the same geometrical process as the resultant of any number of simultaneous velocities.* What is this geometrical process?

154. Review the evidence in favor of the Second Law of Motion.

155. Review the course of reasoning by which the Second Law conducts us to the measure of force.

156. Explain how the Second Law also gives us the means of measuring mass.

157. In what three ways may the quantity g be defined?

158. Show that the dimensions of momentum are $\dfrac{ML}{T}$.

159. Show that the dimensions of the dynamical unit of force are $\dfrac{ML}{T^2}$.

160. A body weighing m lbs. is moved by a constant force which generates in the body in one second a velocity of a feet per second; find the weight which the force could support.

161. Find in what time a force which would support against gravity a mass of 4 lbs. would move a mass of 9 lbs. through 49 feet along a smooth horizontal plane. Find also the velocity acquired.

162. Find how far a force which would support a mass of n units would move a mass of m units in t seconds; and find the velocity acquired.

163. A weight (mass) of 8 lbs. hanging vertically draws a weight (mass) of 12 lbs. by means of a string, passing over a pulley without friction, along a smooth horizontal plane. Find the space described by either body in 4 seconds.

164. If a body weighing 20 kilogrammes be placed on a plane which is made to descend vertically with an acceleration of one metre per second, find the pressure on the plane.

165. If a mass of m units be placed on a plane which is made to ascend vertically with an acceleration f, find the pressure on the plane.

166. A stone, weighing P lbs., is whirled round horizontally by a cord l feet long, having one end fixed; find the time of revolution when the tension of the cord is Q lbs.

LESSON V. — *Statics of a Particle. Parallel Forces. Moments. Equilibrium of a Rigid Body. The Mechanical Powers.*

[In Exercises on the Mechanical Powers, the chief forces concerned are, a *weight* to be supported (or a *pressure* to be exerted), and a *power* by which the weight is supported (or the pressure exerted); these forces are in general denoted by the letters W and P respectively. Unless otherwise stated, friction and the stiffness of cords are neglected, and levers are supposed to be straight, horizontal, perfectly rigid, without weight, and acted upon by parallel forces.]

167. Show how a straight line is employed to denote the magnitude, direction, and point of application of a force.

168. If a force of P units be represented by a straight line a inches long, what force will a straight line f inches long represent?

169. State and illustrate the simple statical Principles or Axioms which may be designated as follows:—

    I. Equal Forces.
    II. Action and Reaction.
    III. Superposition of Forces.
    IV. Transmissibility of Force.
    V. Tension of a Cord.

170. Distinguish between the *direction* and the *line of action* of a force.

171. Show that the resultant of any number of forces

acting along one line is equal to the algebraic sum of the forces. What, then, is the condition of equilibrium for any number of such forces?

172. When two forces act in the same direction they have a resultant of 12 lbs.; when they act in opposite directions their resultant is 2 lbs. Find the forces.

173. Can three forces, represented by the numbers 5, 6, and 12, keep a point at rest?

174. Enunciate the Parallelogram of Forces, the Triangle of Forces, and the Polygon of Forces; and show that all three propositions are deductions from the corresponding propositions relating to velocities.

175. Find the resultant of two forces of 8 lbs. and 15 lbs. respectively, acting on a particle at right angles to each other.

176. What must be the directions of two forces in order that their resultant may be the greatest possible, and also the least possible?

177. Show that as the angle between two forces is increased their resultant will be diminished.

178. Show how to find the resultant of two forces when the directions of the forces and of the resultant, and also the magnitude of one of the forces, are given.

179. Three ropes, PA, QA, RA, are knotted together at the point A; PA is attached to a tree, QA and RA are pulled by two men; given the angle QAR and the force exerted by each man, show how to find the pressure on the tree.

180. Two forces which act at right angles on a point are in the ratio 9 : 40, and their resultant is 123 lbs.; find the forces.

181. The resultant of two forces, P and Q, is perpendicular to P; show that it is less than Q.

182. Resolve 1752 lbs. into two equal forces at right angles.

183. Three equal forces act on a point; find the conditions of equilibrium.

184. A horse tows a boat along a canal, the tow-rope

making an angle of 30° with the course of the boat. If the horse pull with a force of 2000 lbs., find the effective component.

185. Two rafters, making an angle of 120°, support a chandelier weighing 80 lbs.; what will be the pressure along each rafter?

186. A cord is attached to two fixed points, A and B, in the same horizontal line, and bears a ring weighing 10 lbs. at C, so that ACB is a right angle; find the tension of the cord.

187. Forces of 2, 4, 6, and 8 lbs. respectively act along the straight lines drawn from the centre of a square to the angular points taken in order. Find the magnitude and direction of the resultant.

188. Distinguish between *like* and *unlike* parallel forces.

189. Prove that the resultant of two like parallel forces is equal to their sum, and is in the parallel line which divides the distance between the forces into parts inversely as their magnitudes.

190. Prove that the resultant of two unlike parallel forces is equal to their difference, and acts in the direction of the larger force along a line which lies beyond the larger force, reckoning from the smaller force, and the distances of which, from the lines of action of the two forces, are inversely as the magnitudes of the forces.

191. Two men carry a weight of 152 lbs. between them on a pole resting on one shoulder of each; the weight is three times as far from one as from the other. Find how much weight each supports, the weight of the pole being neglected.

192. If P and Q denote two like parallel forces, a the distance apart of their lines of action, and x the distance apart of the lines of action of P and the resultant, prove that

$$x = \frac{Qa}{P+Q}.$$

193. Prove that if any line whatsoever be drawn so as to intersect the lines of action of two parallel forces and their resultant, the portions included between the lines of action of the forces respectively and that of their resultant will be to each other inversely as the forces.

194. Show how to determine by construction the resultant of any number of parallel forces. Is it necessary that the parallel forces should be in one plane?

195. Define the *centre* of a system of parallel forces.

196. Show, from geometrical considerations, that the less the difference in magnitude between two unlike parallel forces, the more remote is the point of application of their resultant. If this difference is nothing, where is the point of application of the resultant?

197. When two unlike parallel forces are equal in magnitude, their resultant, by the general solution, is equal to zero. Show, however, that two such forces are not in equilibrium, and explain what effect they will produce on the body to which they are applied.

198. Define a *couple*, its *arm*, its *moment*, *positive* and *negative* couples.

199. Define the *moment* of a force with respect to a *point*, the moment being the proper numerical measure of the effect of the force to produce rotation around the point. Distinguish between *positive* and *negative* moments.

200. Define the moment of a force with respect to a *line* or *axis*.

201. Prove that the algebraic sum of the moments of three parallel forces in equilibrium round any point in the same plane is equal to zero.

202. Let P_1, P_2, P_3, &c., denote any number of parallel forces acting on a rigid body in one plane, a_1, a_2, a_3, &c., the distances of their lines of action respectively from any point of reference in the plane; and let R denote the resultant of the forces, x its distance from the point of reference: prove that

$$R = P_1 + P_2 + P_3 + \ldots\ldots\ldots + P_n$$
$$R x = P_1 a_1 + P_2 a_2 + P_3 a_3 + \ldots + P_n a_n$$

203. What are the two conditions of equilibrium of

any number of parallel forces in one plane? Examine the cases in which one of these conditions holds true without the other.

204. Define the three kinds of *Levers*. To which kind do the following objects respectively belong, — a *crow-bar*, an *oar*, a *pump handle*, a *common balance*, the *common fire-tongs*, a *pair of nut-crackers*, a *pair of scissors*, the *treadle of a lathe*, the *elbow-joint*.

205. If, in a lever, P and W are the two parallel forces, a and b the *arms*, or perpendicular distances from the fulcrum to the lines of action of the forces respectively, show that the condition of equilibrium is $P a = W b$.

206. If R denote the pressure on the fulcrum of a lever, find R for each of the three kinds of levers.

207. Explain what is meant by *leverage*, or *mechanical advantage* in general, and show in which kind of lever mechanical advantage is always gained, and in which kind it is always lost.

208. If weights of 6 and 9 kilogrammes balance on the ends of a lever 10 metres long, find the position of the fulcrum.

209. On a lever of the second kind a weight of 24 lbs. is suspended at a distance of 8 inches from the fulcrum; the power which holds it in equilibrium is 4 lbs. Find the length of the lever.

210. A lever 8 metres long is supported in a horizontal position by props placed at its extremities; where must a weight of 64 kilogrammes be hung so that the pressure on one of the props shall be 8 kilogrammes?

211. If a body be weighed successively in the two pans of a *false* balance (*i. e.* a balance with unequal arms), prove that the true weight is a mean proportional between the false weights.

212. A man carries a bundle at the end of a stick over his shoulder; as the portion of the stick between his shoulder and his hand is diminished, show that the pressure on his shoulder is increased. Does this change alter his pressure on the ground?

213. At points 5 inches apart, on a rod 20 inches long, weights of 1 lb., 2 lbs., 3 lbs., 4 lbs., and 5 lbs. are suspended. The rod is to be supported in a horizontal position by a single string; at what point must the string be tied?

214. An iron bar 8 feet long rests with its extremities on two props, A and B. Weights of 64 lbs. and 109 lbs. are suspended at points distant 2 feet and 5 feet respectively from A. Find the pressure on each prop.

215. Prove the condition of equilibrium on the *Wheel and Axle*, viz., —

$P \times$ radius of wheel $= W \times$ radius of axle.

216. Find the power required to raise 1 ton by means of a wheel and axle, in which the radius of the wheel is 50 times that of the axle.

217. Prove the condition of equilibrium for the *Single Movable Pulley*, viz. $W = 2P$.

218. Prove the condition of equilibrium for the *First System of Pulleys*, viz. $W = nP$, in which n denotes the number of cords at the lower block.

219. Prove the condition of equilibrium for the *Second System of Pulleys*, viz. $W = 2^n P$, in which n denotes the number of movable pulleys, each hanging by a separate cord.

220. Prove the condition of equilibrium for the *Third System of Pulleys*, viz. $W = (2^n - 1) P$, in which n denotes the number of separate cords, each cord being attached to the weight.

221. In the Second System of Pulleys, $n = 6$, and $P = 28$ lbs.; find W.

222. Find the mechanical advantage obtained by using the Third System of Pulleys with 8 cords.

223. Prove the conditions of equilibrium for the smooth *Inclined Plane*, viz., —

1. $P : W =$ height of plane : length of plane,
2. $R : W =$ base of plane : length of plane,

in which R denotes the reaction of the plane, and P is supposed to act along the length of the plane.

224. The base of an inclined plane is 16 feet, and its height is 12 feet; find the weight which will be supported on the plane by a power of 10 lbs.

225. Find the inclination of an inclined plane if the pressure of the weight on the plane is equal to the power.

226. Prove that if P acts horizontally on an inclined plane, $P : W =$ height of plane : base of plane.

227. Describe the construction of a *Screw*, and show its analogy to an inclined plane.

228. Prove the conditions of equilibrium for the Screw, viz., —

$$\frac{P}{W} = \frac{\text{distance between two threads}}{\text{circumference described by } P}.$$

229. If 5 turns of a screw carry the head forward 1 inch, what power is required to exert a pressure of 1 ton, the length of the lever being 2 feet.

STATICS OF A PARTICLE.

230. Let P and Q denote two forces, R their resultant, and a, β, ϕ, the same angles as in Exercises **88** and **89**; solve the problems of compounding and of resolving two forces by simple substitution in the results of Exercises **88** and **89**, and prove the legitimacy of this method.

231. Taking the formulæ for the composition of two forces, examine the cases, (1) in which $P = Q$ and $a = 120°$, (2) in which $\phi = 0°$, (3) in which $\phi = 180°$, (4) in which $\phi = 90°$.

232. Show that the effective component of a force P along a line which makes an angle a with the line of action of the force is equal to $P \cos a$. Examine the case in which $a = 90°$.

233. Investigate the analytic method of compounding any number of forces acting on a point in one plane, and establish the following formulæ, —

$$X = P_1 \cos a_1 + P_2 \cos a_2 + P_3 \cos a_3 + \ldots + P_n \cos a_n$$
$$Y = P_1 \sin a_1 + P_2 \sin a_2 + P_3 \sin a_3 + \ldots + P_n \sin a_n$$
$$R = \sqrt{X^2 + Y^2}; \tang \phi = \frac{Y}{X}.$$

234. Prove that it is *necessary* and *sufficient* for the equilibrium of any number of forces applied to a point in one plane that the algebraic sums of the resolved parts of the forces in two rectangular directions in the plane be each equal to zero.

92   ELEMENTARY PHYSICS.   [CHAP. I.

235. Four equal forces, each equal to P, act on a point; the first is perpendicular to the second, the resultant of the first two is perpendicular to the third, and the resultant of the first three is perpendicular to the fourth. Find the magnitude of the resultant of all the forces.

236. Four forces, equal respectively to 1 lb., 2 lbs., 3 lbs., and 4 lbs., act on a point; the angle between the directions of the first and third is a right angle, and likewise the angle between the directions of the second and fourth; while the angle between the directions of the first and second is 60°. Find the magnitude and direction of the resultant.

237. Explain how the force of the current may be employed to urge a ferry-boat across a river, the centre of the boat being attached, by means of a long rope, to a mooring in the middle of the stream.

238. What force must a man exert horizontally to push a weight of 336 lbs. a distance of 4 feet from the vertical, supposing it to be suspended from a hook by a rope 20 feet long?

239. The ends of a cord 6 metres long are tied to two hooks in the same horizontal line 3 metres apart, and a smooth ring sliding on the cord sustains a weight of 10 kilogrammes; find the tension of the cord.

240. From the highest point, C, common to two equal rafters, CA and CB, is suspended a chandelier, the weight of which is W; find the horizontal *thrust* at A or B.

241. A man with a perfectly smooth spherical head wears a conical hat; find the total pressure on his head. What remarkable consequence follows, if we suppose the hat to be very high, its weight remaining the same?

242. Find the resultant of three forces acting on a point in lines at right angles to one another.

243. Forces act on a point in any direction; find the magnitude and direction of their resultant.

244. What are the conditions of equilibrium of a particle under the action of any system of forces?

MOMENTS AND COUPLES.

245. Prove that if any number of forces act on a point in one plane the algebraic sum of their moments about any point in their plane is equal to the moment of their resultant about the same point. In what two cases is this algebraic sum equal to zero?

246. Define the *axis* of a couple.

247. Show that a couple cannot be balanced by a single force, but that it can be balanced by another couple.

248. Show that the effect of a force at a point not in its line of action is equivalent to an equal force applied at that point, together with a couple formed of the original force and an equal force supposed to be applied to the point in the opposite direction.

249. Prove that three forces which act in consecutive directions round a triangle, and are represented respectively by its sides, are not in equilibrium, but are equivalent to a couple.

250. Prove that any two couples in the same plane, if they have equal moments with unlike signs, balance one another.

EQUILIBRIUM OF FORCES IN ONE PLANE.

251. Show that any system of forces acting in one plane on a rigid body can be reduced to a single force or to a couple.

252. Demonstrate the necessary and sufficient conditions of equilibrium of any number of forces acting in one plane on a rigid body, viz., —

1°. *The algebraic sums of the projections of the forces, estimated along any two rectangular axes in the same plane as the forces, must each be equal to zero.*

2°. *The algebraic sum of the moments of the forces with respect to any point in the plane of the forces must be equal to zero.*

253. At what point of a tree must a rope of a given length (l) be attached, in order that a man on the ground, pulling at the other end, may produce the greatest effect in overturning the tree?

254. A sphere, the weight of which is W, rests on two planes inclined at angles α and β to the horizon; find the normal pressures on the planes.

255. A hemispherical bowl in a fixed position, with its rim horizontal, contains a weight (W) which is attached to a weight (P) outside the bowl by means of a string passing over a pulley on the rim of the bowl. Find the position of equilibrium of the weight in the bowl, friction being neglected.

[In the next four Exercises the weight of a material body is to be considered as another force applied at the *centre of gravity* of the body, and acting vertically downwards (see Lesson VII.).]

256. A roof is composed of equal beams, joined in pairs, forming the sides of isosceles triangles; find the horizontal thrust on the side walls.

257. Two beams of known lengths, m and n, connected together at a given angle ϕ, turn about a horizontal axis at their point of junction; find the position of equilibrium which they will take by their own weight.

258. A uniform beam rests upon a rail, with one extremity in contact with a smooth vertical wall; find the conditions of equilibrium.

259. A ladder rests on a smooth floor with its upper end leaning against a smooth vertical wall; determine the least horizontal force which, applied at the lower end of the ladder, will prevent it from slipping. Find also how this force is altered as a man ascends the ladder.

THE MECHANICAL POWERS.

260. The weight (W) being given in each of the three kinds of levers, show within what limits the magnitude of the pressure on the fulcrum will lie.

261. Find the condition of equilibrium in a combination of levers or *compound* lever.

262. A bent lever has equal arms, making an angle of 120°; find the ratio of the weights at the ends of the arms when the lever is in equilibrium, with one arm horizontal.

263. Prove that in a combination of Wheels and Axles,

$$\frac{P}{W} = \frac{\text{product of radii of the axles}}{\text{product of radii of the wheels}}.$$

264. A Wheel and Axle is applied to sustain a weight on an inclined plane, the string being parallel to the plane; find the conditions of equilibrium.

265. Prove that the mechanical advantage of the single movable pulley will be diminished by taking into account the weight of the pulley.

266. Find, by the aid of trigonometry, the conditions of equilibrium on the Inclined Plane, viz., —

$$P = W \sin a \; ; \; R = W \cos a :$$

and show that these conditions are identical with those given in Exercise **223**.

267. If the force required to draw a wagon on a horizontal road be $\frac{1}{70}$ of the weight of the wagon, what will be the force required to draw it up a hill, the slope of which is 1 in 40?

268. Two unequal weights are placed on two smooth inclined planes having a common height, and are connected by a fine string passing over the intersection of the planes; find the ratio between the weights when there is equilibrium.

269. The length of the power-arm in a Screw is 15 inches; find the distance between two threads of the Screw in order that the mechanical advantage may be 30.

270. The pitch of a screw is 30°, the radius of the cylinder .9 inches, and the length of the arm 4 feet; find the power that will exert a pressure of 1 ton.

LESS. VI.] *EXERCISES AND PROBLEMS.* 95

LESSON VI. — *Third Law of Motion. Impact.*

271. State the Third Law of Motion in the terms employed by Newton, and give his own illustrations.

272. Distinguish between *exterior* and *interior* forces, and illustrate the distinction by an example.

273. What illustration of the Third Law is afforded in leaping from a small boat to the shore?

274. If a nail is to be driven into a board which is not firmly supported, it enters much better provided we place behind the board some solid body, as a block of iron. Explain this.

275. A boat is 100 yards distant from a ship, and a man in the boat hauls the boat to the ship's side by means of a rope extending from the ship's side to the boat. The masses of the boat and ship are 200 lbs. and 1600 tons respectively. Find where the ship and boat meet, supposing no difference in the resistance of the water.

276. A gun weighing 5 tons is charged with a ball weighing 28 lbs.; if the gun be free to move, with what velocity will it recoil when the ball leaves it with a velocity of 1000 feet per second?

277. It is related of a gentleman that he thought he had found the means of commanding at any time a fair wind for his pleasure-boat by placing an immense bellows in the stern, blowing against the sail. Point out the uselessness of such an arrangement.

278. A man is placed on a perfectly smooth table; show how he may get off.

279. Distinguish between *impulsive* and *continuous* forces, and give illustrations. What is the measure of an impulsive force?

280. Distinguish between *perfectly elastic, imperfectly elastic,* and *perfectly inelastic* bodies. Define the *coefficient of restitution.*

281. State the principle of the *Conservation of Momentum,* and explain its application to the case in which two equal bodies moving with equal velocities in opposite directions impinge on one another.

282. Two perfectly inelastic balls, moving along the same line with given velocities, impinge directly upon each other ; find the common velocity after impact.

283. An inelastic body, weighing 250 kilogrammes and moving with a velocity of 20 metres per second, meets another inelastic body weighing 300 kilogrammes and moving with a velocity of 2 metres per second in the opposite direction. Find the common velocity after impact.

284. Three inelastic balls, weighing respectively 5 lbs., 7 lbs., and 8 lbs., lie in the same straight line. The first is made to impinge on the second with a velocity of 60 feet per second ; the first and second together then impinge on the third. Find the final velocity.

285. What fact is overlooked in the following objection to the Third Law ? "When an obstacle gives way under a force applied to it, the reaction must be less than the action; for how otherwise can the yielding of the obstacle be explained than by the consideration that a greater force overcomes a less ?"

286. Illustrate the Third Law in the case of a horse and cart, showing what force or forces are balanced by the effort of the horse, (1) when the motion is uniform ; (2) when horse and cart are starting ; (3) when the cart is stopping, the horse still pulling; (4) when the cart is stopping, the horse being backed.

287. Distinguish between *impressed* and *effective* forces.

288. State and explain D'Alembert's Principle.

289. Two heavy bodies are connected by an inextensible string without weight, which passes over a fixed smooth pulley; determine the motion.

290. If two unequal weights, connected by a string, be allowed to fall, the string being vertical, what will be the tension of the string ?

291. Determine the motions of two weights, W, W', along inclined planes, placed back to back, the weights being connected by a thread.

292. Demonstrate the principle of the Conservation of Momentum, stated in a general form as follows: *the sum of the momenta of the parts of any material system, estimated in any direction, is unchanged by the action of interior forces.*

293. Demonstrate the formulæ for the impact of two imperfectly elastic bodies, viz., —

$$(m + n)\, u' = m\, u + n\, v - e\, n\, (u - v),$$
$$(m + n)\, v' = m\, u + n\, v + e\, m\, (u - v),$$

LESS. VI.]   *EXERCISES AND PROBLEMS.*   97

in which e is the coefficient of restitution, m and n the masses of the bodies, u and v the velocities before impact, u' and v' the velocities after impact.

294. Show that the formulæ of the preceding Exercise include the cases of perfectly elastic and perfectly inelastic bodies, and adapt the formulæ to these cases.

295. Prove that the relative velocities of two perfectly elastic bodies before and after direct impact are equal and opposite.

296. Define *vis viva*, and prove that in the impact of perfectly elastic balls no *vis viva* is lost.

297. Prove that in the impact of imperfectly elastic balls *vis viva* is lost, and find what proportion of the whole *vis viva* is lost.

298. A ball impinges on a fixed plane with a velocity given in direction and magnitude; the value of the coefficient of restitution between the ball and the plane is e; find the motion of the ball after impact, friction being neglected.

299. A billiard-ball is struck from one corner, A, of a billiard-table, $ABCD$, and after striking three of its sides falls into the pocket at B; show that the alternate sides of its course are parallel, and find the distance of the first point of impact from B, if $AB = a$, and $BC = b$.

300. Review the evidence in favor of the Third Law of Motion.

CHAPTER II.

THE FORCES OF NATURE.

LESSON VII. — *Universal Gravitation.*

301. When a pendulum is pulled aside from its vertical position and then left free, why does it not swing to an equal distance on the other side of the vertical position ?

302. Newton, in his pendulum experiments, employed as the bob of his pendulum a box into which he put different substances equal in weight. What reason was there for employing the box ?

303. Comment on the sentence on page 42 of Mr. Stewart's book, beginning, " But had the masses been different in Newton's experiment.". (See top of page 14.)

304. What is the angle between the directions of plumblines at the north pole and at the equator ?

305. State the proposition referred to on page 44 of Mr. Stewart's book as " a well-known proposition in geometry."

306. State Newton's Law of Universal Gravitation in the form of an algebraic formula, and derive from the formula the simplest unit of attractive force.

307. The attractive force between two masses is 4, and the values of the masses are 20 and 5 ; find their distance apart.

308. Divide a given mass into two parts, such that the mutual attraction of the parts may be a maximum.

309. Explain precisely how far the Law of Gravitation is established by the proof given in § 37 of Mr. Stewart's book.
310. State Kepler's Laws. How were they established ?
311. The areas which revolving bodies describe by radii drawn to an immovable centre of force lie in the same immovable planes and are proportional to the times in which they are described. (*Newton.*)

[LESS. VII.] *EXERCISES AND PROBLEMS.* 99

312. Every body that moves in any curved line described in a plane, in such a manner that a radius drawn from the body to a point in the plane, either fixed or in a state of uniform rectilinear motion, describes about the point areas proportional to the times, is urged by a centripetal force directed to that point. (*Newton*.)

313. If a body describe areas proportional to the times about another body, however moved, the force which acts on the first body is the resultant of two forces, — a central force towards the second body, and an accelerating force common to both bodies. (*Newton*.)

314. The velocity of a body attracted towards an immovable centre, in spaces void of resistance, is inversely as the perpendicular let fall from that centre on the tangent to the path. (*Newton*.)

315. The path of a body being an ellipse, and the centre of force being at one of the foci, it is required to find the law of the force. (*Newton*.)

316. Give the dynamical interpretation of Kepler's Laws, and their relations to the Law of Universal Gravitation.

317. Show that Kepler's Third Law, taken in connection with Newton's Third Law of Motion, leads to the conclusion that the mutual attraction between the sun and a planet must be directly proportional to the mass of the sun as well as to that of the planet.

318. Two bodies attract one another according to the Law of Gravitation; determine the actual character of the motion.

319. Prove that the accurate statement of Kepler's Third Law is

$$T^2 : T'^2 = \frac{R^3}{M+m} : \frac{R'^3}{M+m'}$$

in which M denotes the mass of the sun, m and m' the masses of any two planets, T and T' their periods, and R and R' their mean distances from the sun. What inference may be drawn as to the masses of the planets?

320. Give a summary of the evidence in favor of the Law of Universal Gravitation.

321. Investigate the effect of the centrifugal force due to the earth's daily rotation upon the weight of a body at the equator.

322. Find the time of revolution of the earth, which would cause bodies to have no weight at the equator.

323. Investigate the effect of centrifugal force at any place upon the weight of a body at that place.

324. Explain how the centrifugal force due to rotation tends to alter the form of the rotating body.

325. Explain the total variation which is known to exist between the force of gravity at the equator and at the pole.

100 ELEMENTARY PHYSICS. [CHAP. II.

LESSON VIII. — *Attwood's Machine.*

326. Why is it that a ball of lead and another of cork, precisely equal in volume, will not fall to the earth with equal velocities ?

327. In Attwood's Machine, what reason is there for having the wheels at the top as small as possible ?

328. Solve the following general exercise on Attwood's Machine : Two heavy bodies are connected by an inextensible string, which passes over a fixed pulley ; find the tension of the string and the motions of the bodies.

329. In Attwood's Machine, given $P = 12$ oz., $Q = 6$ oz.; find the pressure on the axis of the pulley.

330. In Attwood's Machine, given $P = Q = 18.6$ oz.; what weight must be added to P in order that it may descend through 100 ft. in 8 seconds ?

331. In *Experiment H* with Attwood's Machine (see Stewart, pp. 51, 52), the boxes are observed to rise and fall alternately through diminishing distances for some time ; explain this.

332. A weight of two pounds hanging vertically draws another weight of three pounds up a smooth plane inclined at an angle of 30° to the horizon ; find the space described in 4 seconds.

333. Two scales are suspended by a string over a small pulley ; six equal bullets are placed in one scale and six in the other ; show that the tension of the string is greater with this arrangement of the bullets than with any other.

334. If two unequal weights connected by a string be allowed to fall, the string being vertical, what will be the tension of the string ?

LESSON IX. — *Centre of Gravity. Pendulum.*

335. Give an exact definition of the centre of gravity of a body.

336. Define and illustrate the terms *homogeneous body, plane of symmetry, centre of figure* or *geometric centre.*

337. Where is the centre of gravity of a homogeneous body, having planes of symmetry ?

[LESS. IX.] EXERCISES AND PROBLEMS.

Locate the centre of gravity in the following cases : *A straight line, circumference of a circle, area of a circle, area of a parallelogram, surface of a sphere, volume of a sphere, convex surface of a right cylinder, volume of a right cylinder, volume of a parallelopiped.*

338. Show how to find the centre of gravity of any number of heavy points.

339. Weights of 1, 2, and 3 pounds are placed along the same line a foot apart; find their centre of gravity.

340. Examine the equilibrium of a body suspended from a fixed point.

341. Examine the equilibrium of a body resting upon a fixed point.

342. Examine the equilibrium of a body resting upon two fixed points.

343. Examine the equilibrium of a body resting on three or more fixed points.

344. Examine the equilibrium of a right cylinder resting on a horizontal table, (1) on its base, (2) on its side.

345. Explain why a load of hay will overset on the side of a hill, when a load of iron of equal weight will pass along in safety.

346. Explain why it is difficult, if not impossible, for a person standing with his heels against a wall, to pick up a cent between his feet.

347. A balance is in equilibrium with horizontal beam, unloaded pans, and the points of suspension of the pans at a higher level than the fulcrum; explain the effect produced by loading the pans.

348. Show how to determine the value of the acceleration of gravity at any place by pendulum observations.

349. Given the length of a second's pendulum, find the length of a pendulum which will oscillate once a minute.

350. If a pendulum were taken to a place where the force of gravity was increased fourfold, how much would the length of the pendulum have to be changed in order that the time of oscillation should remain the same as before?

351. A pendulum at Paris one metre long was found to oscillate in 1·00304 seconds ; find the value of g at Paris.

352. Show how to find the centre of gravity of a triangle.

353. A uniform bar, 4 feet long, weighs 10 lbs.; and weights of 30 lbs. and 40 lbs. are placed on its two extremities; on what point will it balance ?

354. How long a piece must be cut off from one end of a rod of length $2\,a$, in order that the centre of gravity of the rod may approach towards the other end through a distance b ?

355. A beetle crawls from one end of a straight fixed rod to the other end; find the consequent alteration in the position of the centre of gravity of the rod and beetle.

356. Two homogeneous spheres of equal density touch each other; find the distance of their centre of gravity from the point of contact, the radii being respectively 8 inches and 12 inches.

357. How high can a cylindrically shaped tower of r metres radius be built without falling, if it be inclined from the vertical by an angle of θ degrees ?

358. Show how the requisites of a good balance may be satisfied.

359. A shopkeeper has correct weights, but a false balance; supposing that he serves out to two customers articles weighing W lbs. by his balance, using first one scale and then the other, find whether he gains or loses on the whole, and how much.

360. Prove that the time of one small oscillation of a simple pendulum of length l is equal to

$$\pi \sqrt{\frac{l}{g}}$$

361. A pendulum, which would oscillate once a second at the equator, would gain 5 minutes a day at the pole; compare equatorial and polar gravity.

362. If two pendulums, of lengths l and l', at different points on the earth's surface make in the same time numbers of vibrations which are in the ratio $m : m'$; find the ratio between the forces of gravity at the two places.

363. Show how the height of a mountain may be ascertained by pendulum observations.

364. A seconds pendulum is taken to the top of a mountain of height h; find the number of seconds it will lose in one day.

365. A seconds pendulum, on being taken to the bottom of a mine, was found to lose 10 seconds a day; find the depth of the mine, given that the earth's radius is equal to 4000 miles, and that *in the interior of the earth gravity varies directly as the distance from the earth's centre*.

366. A pendulum, when taken to the top of a mountain, is observed to lose daily just twice as much as it does when taken to the bottom of a mine in the neighborhood; show that the height of the mountain is equal to the depth of the mine.

367. The times of oscillation of a pendulum are observed on the earth's surface and at the bottom of a mine; hence find the radius of the earth, supposed spherical.

LESSON X. — *Forces exhibited in Solids*.

368. Define and illustrate the terms *coefficient of friction*, *angle of friction*.

369. What are the forces acting on a body which stands at rest on the side of a hill?

370. What are the forces acting on a ladder which stands with one end on a rough horizontal floor and the other end resting against a rough vertical wall?

371. Prove that the coefficient of friction is equal to the tangent of the angle of friction (or to the ratio of the height to the base of an inclined plane the angle of inclination of which is equal to the angle of friction).

372. A body will just rest on a plane inclined at an angle of 45°; find the coefficient of friction.

373. The height of a rough inclined plane is to the length as 3 : 5, and a weight of 30 lbs. is just supported on the plane by the friction; find (1) the force of friction, (2) the coefficient of friction.

374. Compare the strength of two beams, one of which is twice as long and twice as deep as the other, their breadths being the same.

375. Since, of beams of the same section, the deeper is the stronger, why are not beams made in practice exceedingly thin and deep?

Lesson XI. — *Forces exhibited in Liquids.*

[In the exercises on this Lesson the pressure of the atmosphere is neglected. For specific gravities, see Appendix V.]

376. Define a *perfect* fluid. Define and illustrate *viscosity*.

377. State the chief differences between a *solid* and a *fluid*; between a *liquid* and a *gas*.

378. What necessary consequence as to the direction of fluid pressure follows immediately from the definition of a perfect fluid?

379. State Pascal's principle in a mathematical form. Why cannot the principle be completely established by direct experiment?

380. Show that any force, however small, may, by transmission through a fluid, be made to support any weight, however large.

381. Show how, by the weight of a few ounces of water, the strongest cask may be burst.

382. Prove that in Bramah's press we have a direct verification of the general principle of Mechanics, that *what is gained in force is lost in velocity*.

383. A vessel full of liquid has two pistons, 3 and 18 centimetres in diameter respectively; what pressure on the smaller will produce a pressure of 900 kilogrammes on the larger?

384. A closed vessel full of liquid has a weak part in its upper surface, not able to bear a pressure greater than 9 lbs. per square foot. If a piston the area of which is one square inch be fitted into an opening in the upper surface, what pressure applied to it will burst the vessel?

385. Prove that *the free surface of a liquid at rest is normal at every point to the resultant of all the forces acting at that point.*

386. Explain the form of the free surface of a liquid at rest on the surface of the earth.

387. If we consider any horizontal plane in a liquid at rest under the action of gravity, the pressure on the plane is proportional, (1) to the area of the plane, (2) to the depth of the plane, (3) to the density of the liquid. On what grounds do these laws rest ?

388. Prove that the pressure on any horizontal surface in a liquid is given by the formula, —

$$Pressure\ (in\ grammes) = \begin{cases} area\ (in\ c.\ m.^2) \times depth\ (in\ c.\ m.) \\ \times density. \end{cases}$$

389. Prove that the pressure on any vertical surface in a liquid is given by the formula

$$P = SHD$$

where P denotes the pressure, H the depth of the centre of gravity of the surface (supposed uniform in density), and D the density of the liquid. Specify the units in which P, S, H, and D respectively should be expressed. The pressure is equal to the weight of what column of water ?

390. Find the pressure at the depth of 30 metres in a lake.

391. What height must a column of water have which will exert a pressure of 1000 kilogrammes per square decimetre ?

392. Required the pressure on a rectangular vertical side of a tank full of water, the height of the tank being 4 metres, and its breadth being 80 centimetres.

393. Required the pressure on a vertical triangle immersed in water with the base in the surface, the base being 50 centimetres, and the altitude 30 centimetres.

394. Prove that the total pressure experienced by a cubical vessel full of water is equal to three times the weight of the water.

395. Two cubical vessels, the edges of which are as two to one, are filled with water ; compare the pressures, (1) on their bases, (2) on their total interior surfaces.

396. Sketch the form of a vessel in which the pressure

against the sides shall much exceed the pressure on the base.

397. Taking the density of mercury as 13·6, find the total interior pressure in a cylinder full of mercury, the height of the cylinder being one metre, and the radius of the base being 16 centimetres.

39̂. Prove that the common surface of two liquids which do not mix must be horizontal.

399. Prove that the heights of two vertical liquid columns in communication are inversely as the densities of the liquids.

400. Distinguish between the *centre of gravity* and the *centre of pressure* of a vertical surface pressed by a liquid.

401. How will the pressure on the base of a vessel containing water be affected by dipping a piece of metal into the water, (1) when the vessel is just full of water, (2) when the vessel is not full?

402. Prove that when a body is placed in a liquid, it will (1) sink, (2) remain at rest, or (3) rise to the surface of the liquid, according as its density (or specific gravity) is (1) greater than, (2) equal to, or (3) less than that of the liquid.

403. Find the weight of a boat which displaces 10 cubic metres of water.

404. Prove that, when a body floats on the surface of a liquid, the *part of the body immersed* is to *the whole body* as the *density of the body* is to *that of the liquid*.

405. Define and distinguish between *density* and *specific gravity*, and explain how each is measured.

406. A body measuring 18 cubic centimetres floats in water with its whole bulk immersed; find its weight.

407. Find the weight in water of 100 cubic centimetres of iron.

408. Find the weight of one cubic centimetre of a body which floats in water with one fifth of its volume above the surface.

409. An irregularly shaped mass of granite weighs 182 kilogrammes in air and 117 kilogrammes in water; find its volume and its specific gravity.

410. Two bodies differing in bulk weigh the same in water. Which will weigh the most in mercury and which the most in a vacuum?

411. A man (specific gravity 1·12) weighs 70 kilogrammes. What volume of cork will be required to just float him in water?

412. A cylinder of wood floats in water with the axis vertical. How much will it be depressed by putting a weight W on top of it?

413. Suppose that a man exerting all his strength can just raise a weight of 100 kilogrammes; find the weight of a stone (specific gravity 2·5) which he can raise under water.

414. Show how to find the specific gravity of a mixture of given volumes of any number of given substances.

415. Show how to find the specific gravity of a mixture of given weights of any number of given substances.

416. Find the specific gravity of a mixture of equal volumes of water and alcohol, supposing no contraction to take place.

417. Find the specific gravity of a mixture of equal weights of water and alcohol, supposing no contraction to take place.

418. Three liquids, the specific gravities of which are respectively 1·2, 0·96, and 1·456, are mixed in the proportions by volume of 18 parts of the first to 16 parts of the second and 15 parts of the third. Find the specific gravity of the mixture.

419. What are the proportions of gold and silver in an alloy of these two metals which weighs 10 kilogrammes in air and 9·375 kilogrammes in water?

420. If a diamond ring weighs 69·5 grammes in air and 64·5 grammes in water, find the weight of the diamond in the ring.

421. Find the specific gravity of a piece of lead which weighs 47·48 grammes in air and 43·33 grammes in water.

422. Explain a method of finding the specific gravity of a solid lighter than water.

423. A block of wood, weighing in air 8 lbs., is tied to a piece of metal weighing 6 lbs.; in water both together weigh 4 lbs., while the metal alone weighs 5 lbs. Find the specific gravity of the wood.

424. A crystal of salt weighs 6·3 grammes in air; when covered with wax, the whole weighs 8·22 grammes in air, and 3·02 grammes in water; find the specific gravity of the salt.

425. A glass ball, weighing 10 grammes in air, loses 3.636 grammes in water, and 2·88 grammes in alcohol. What is the specific gravity of the alcohol?

426. Prove that the volumes of different liquids displaced by the same floating body are inversely as the specific gravities of the liquids.

How, by this principle, can the specific gravity of a liquid be ascertained?

427. Prove that the entire interior pressure on a hollow sphere full of a liquid is equal to three times the weight of the liquid.

428. A rectangle is just immersed vertically in a liquid, with one side in the surface; divide it by a horizontal line into two parts which shall be equally pressed by the liquid.

429. Divide a rectangle just immersed vertically in a liquid, with one side in the surface, by horizontal lines into n parts on which the pressures shall be equal.

430. Distinguish between *total* pressure and *resultant* pressure by the aid of illustrations (as a cubical vessel full of water, a teacup full of tea, &c.), and prove that when a vessel of any shape is filled with a liquid the resultant pressure is equal to the weight of the liquid.

431. A hollow cylinder (height h, radius of base r) is filled with a liquid; compare the total pressure on its interior surface with the weight of the liquid.

432. A right cone (altitude h, radius of base r) rests on its base and is full of water; compare the entire interior pressure with the weight of the water.

433. Prove that the centre of pressure of a rectangular vertical area is on the vertical line which passes through the centre of gravity of the area at a distance of two thirds of the height of the area from the surface of the liquid.

434. Prove that, as a plane area is lowered vertically in a liquid, the centre of pressure approaches, and ultimately coincides with, the centre of gravity of the area.

LESS. XI.] *EXERCISES AND PROBLEMS.* 109

435. State the general conditions of equilibrium of a floating body, and define the states of stable, unstable, and neutral equilibrium, with illustrations.

436. Define the *metacentre* of a floating body and prove that the equilibrium of a floating body is *stable*, *unstable*, or *neutral*, according as the metacentre is *above*, *below*, or *at* the centre of gravity of the body.

437. A ship sailing into a river sinks 2 centimetres, and after discharging 12000 kilogrammes of her cargo, rises 1 centimetre; determine the weight of ship and cargo, the specific gravity of seawater being 1.026.

438. What quantities of zinc and copper must be taken to make an alloy weighing 50 grammes, and the specific gravity of which shall be 8·2?

439. Given the weights and specific gravities of two bodies of different kind; find the specific gravity of the compound formed by mixing them, (1) when the contraction is $\frac{1}{m}$th part of the sum of the component volumes, (2) when the expansion is $\frac{1}{n}$th part of the sum of the component volumes.

440. Required the specific gravity of a mixture of 18 kilogrammes of sulphuric acid and 8 kilogrammes of water, assuming that the contraction is $\frac{1}{32}$.

441. Three masses of gold, silver, and a compound of gold and silver, weigh respectively, P, Q, and R grammes in air, and p, q, and r grammes in water; find the weight of gold in the compound.

442. The weight W of a vessel full of air (specific gravity ρ) and the weight W of the same vessel full of a liquid (specific gravity σ) being given; find the capacity of the vessel.

443. Given the apparent weight P of a body in air, and its specific gravity σ; find its true weight (or weight *in vacuo*).

444. A cube of lead (edge 4 centimetres) is to be sustained under water by attaching to it a sphere of cork; required the diameter of the sphere of cork which will just sustain it.

445. A mass of copper is suspected of being hollow. Its weight in air is 523 grammes, in water 447·5 grammes. Find the volume of the interior cavity.

446. A piece of wood (specific gravity 0·729) is in the form of a right cone, and is floating in water with its axis vertical; determine how much of the height of the cone will be submerged, (1) when the vertex is below, (2) when the vertex is above.

447. An iron cone is floating on mercury with its vertex downwards; required the ratio of the altitude of the cone immersed to the total altitude of the cone.

448. Determine the ratio of the thickness of a hollow iron globe to its diameter in order that it may just float in water.

449. A sphere of cork is set free at the depth of 80 feet in a lake; in what time will it reach the surface of the lake, supposing that it experiences no resistance to motion from the water?

450. Determine the general relation between the volumes of liquids displaced by a hydrometer of variable immersion and the specific gravities of the liquids.

451. Explain the graphical method of graduating a hydrometer.

LESSON XII. — *Forces exhibited in Gases.*

[The specific gravities of gases are usually referred to that of dry air at the same pressure and temperature as a standard. See Appendix V., Table of Specific Gravities.]

452. Describe three forms of expressing the pressure or tension of a gaseous body which are in use.

453. Find the pressure of the atmosphere on the surface of a glass globe, 20 centimetres in diameter, when the barometer stands at 76 centimetres.

454. The body of a man of average size exposes a surface of about 1·5 square metres; find the total atmospheric pressure upon it when the barometer stands at 72 centimetres.

455. What is the pressure of the atmosphere per unit of area in dynamical measure, when the height of the barometer is h, and in a locality where the force of gravity on unit of mass is g?

456. Required the height of a water barometer when the mercury barometer stands at 76 centimetres.

457. Find the height of an alcohol barometer, when that of the water barometer is 33 feet.

458. A barometer is partly filled with water and partly with mercury, the height of the water being three times that of the mercury; find the total height when the pressure of the atmosphere is 1020 grammes per square centimetre.

459. A barometric tube less than 30 inches in length

was filled with mercury, and then weighed. It was then inverted in the usual way over a basin of mercury, attached to the beam of a balance so as to be in a vertical position, and again weighed. It was found to weigh precisely the same as before. Explain this.

460. A weight, suspended by a string from a fixed point, is partially immersed in water; will the tension of the string be increased or diminished as the barometer rises?

461. If a piece of glass float on the mercury within a barometer, will the mercury stand higher or lower in consequence?

462. Explain how depths below the sea-level may be determined by means of barometric observations in a diving-bell.

463. If the height of a barometer in a diving-bell is 200 inches, the height at the sea-level being 30 inches, find the depth descended.

464. Prove that the height of the atmosphere supposed *homogeneous* (*i. e.* of the same density throughout as at the surface of the sea) would be about 7987 metres, or rather less than 5 miles.

465. Deduce a formula for finding the ascensional force of a spherical balloon.

466. Calculate the ascensional force of a spherical soap-bubble, 10 centimetres in diameter, filled with coal-gas at a tension of 78 centimetres, the weight of the watery film being neglected.

467. Prove that the specific gravity of a gas with reference to water is equal to the product of its specific gravity with reference to air and the specific gravity of air with reference to water.

468. Give a mathematical statement of Boyle's Law, supposed absolutely true. Give a mathematical statement of this law which includes the ascertained deviations from the law.

469. Prove that the specific gravity (and also the mass of a given volume) of a gas varies directly as the pressure applied to it.

470. A cubic foot of air is compressed into a cubic inch; find the pressure required.

471. A vessel with elastic sides contains 7·545 litres of air under a pressure of 64 centimetres; required the volume under the standard pressure of 76 centimetres.

472. If a small hole be made in the top of a submerged diving-bell, will the air flow out or the water flow in?

473. If a block of wood be floating on the surface of the water within a diving-bell, how will it be affected by the descent of the bell?

474. How would the tension of the rope which sustains a diving-bell be affected by opening a bottle of soda-water in the bell?

475. Suppose a cylindrical diving-bell is lowered under the sea by means of a rope; show that, unless air is forced in from above, the tension of the rope will increase as the bell descends.

476. Supposing a diving-bell cylindrical, and that no air is supplied from above; find the height to which the water will rise in the bell for a given depth descended.

477. If a cylindrical tube 152 centimetres long be half filled with mercury and then inverted over mercury in a basin, determine how high the mercury will stand in the tube when the height in a perfect barometer is 76 centimetres.

478. Deduce a formula for expressing the rarefaction produced in an air-pump by n strokes of the piston.

479. What causes prevent us from obtaining a perfect vacuum with an air-pump?

480. Show that the single-barrelled air-pump works harder as the rarefaction proceeds.

481. If the capacity of the receiver of an air-pump be three times that of the barrel, what will be the pressure of the air in the receiver after three strokes of the piston, the initial pressure being 15 pounds per square inch?

482. If the tension of the air in the receiver of an air-pump is reduced to one fourth its original amount by

LESS. XII.] *EXERCISES AND PROBLEMS.* 113

three strokes of the piston, compare the capacities of the receiver and barrel.

483. Find the force required to raise the piston of the common pump, neglecting friction and the weight of the piston.

484. Prove that the work done in raising a given quantity of water by a pump is the same as would be done in raising the water through an equal distance in buckets, friction and other hurtful resistances being neglected.

485. How high can mercury be raised by a common suction-pump?

486. When a siphon is working, what would be the effect of making a small hole at its highest point?

487. What effect upon the action of a siphon would be produced by carrying it up a mountain?

488. What effect would be produced on the action of a siphon, if the atmosphere were suddenly to become denser than water?

489. The common radius of two Magdeburg hemispheres is 16 centimetres, and the air is exhausted until the interior pressure is reduced to 4 centimetres; required the force necessary to separate the hemispheres, when the height of the barometer is 76 centimetres.

490. Prove that the weight in statical or gravitation measure of a litre of dry air at 76 centimetres pressure in any locality is proportional to the intensity of the force of gravity at that locality; and that its weight in dynamical measure is proportional to the square of the intensity of the force of gravity, temperature being supposed constant.

491. Prove that, as the heights above the earth's surface increase in arithmetical progression, the densities of the air decrease in geometrical progression, temperature, moisture, and intensity of gravity being supposed uniform.

492. Deduce the formulæ given in Appendix V. for ascertaining the height of a mountain by barometric observations.

493. Deduce the formulæ given in Appendix V. for determining the ascensional force of a balloon.

494. A spherical air-bubble ascends in water; given its diameter at the depth a, find its diameter at the depth $\frac{a}{4}$.

495. Explain how to graduate the compressed-air manometer, the tube being supposed uniform in diameter.

H

496. A small quantity of air is left in the upper part of a barometric tube ; determine the effect on the height of the mercury column.

497. The weights of a body in air are a, a', corresponding to the heights h, h', of the barometer ; find the weight corresponding to a height h''.

498. Find the limit to the exhaustion of air by an air-pump due to the fact that the piston does not descend to the bottom of the barrel (*i. e.* to the existence of what may be termed *untraversed space*).

499. Determine the law of the rarefaction of air in an air-pump, taking into account the existence of untraversed space.

500. Find the law of the increase of the tension of the air in the receiver of a condensing-pump.

501. The capacity of the receiver of a condensing-pump is 10 litres, and that of the barrel is 200 cubic centimetres ; how many strokes are required to bring the tension of the air in the receiver to 8 atmospheres ?

502. Investigate the conditions under which the suction-pump will not work, when the piston does not descend to the fixed valve.

PART III.

ANSWERS AND SOLUTIONS.

I. — ANSWERS.

Lesson I.

10. 5401440 square feet.
11. 1240000 sq. decimetres.
12. 7415·375 cubic yards.
13. 346·768595 cubic metres.
14. 2000 litres.
15. 6 tonnes.
16. 14·174765 grammes.
17. As 1 to 1000000.
18. 64 kilogrammes; 64 grms.
20. 19200 kilogrammes.
24. 17·76 kilogrammes.
27. 3 miles per hour.
28. As 2 to 1.
29. $1527\frac{1}{3}$ feet per second.
30. 39 feet per second.
31. 13 feet per second.
32. As π to 1.
33. 122522·1114 kilogrammes.
34. 714·28 cubic centimetres.
35. 0·54.

37. The density of every substance would be increased $\dfrac{b}{a}$ times.

38. The density of the wood is $\dfrac{10\,c}{\pi\,b^2\,d}$ times that of the substance whose density is d.

39. $\dfrac{M}{L^3}$, M and L representing in general a mass and a length respectively.

41. The linear velocity must be equal to the radius of the circle.

42. $\dfrac{\pi}{21600}$; $\dfrac{\pi}{1800}$.

44. (1) $a + (v' - v)\,t$; (2) $\dfrac{a}{v - v'}$; (3) $\dfrac{a\,v}{v - v'}$. **45.** 360.

ELEMENTARY PHYSICS.

Lesson III.

57. (1) The direction of the diagonal of a parallelogram the sides of which are 4 in the direction of the rowing, and 3 in the direction of the current. (2) 2·5 miles. (3) 1·5 miles. (4) Half an hour. (5) Half an hour.

63. $8\sqrt{2}$ acting towards the northwest.

64. The ball would be at rest relatively to the ground.

68. (1) $36\frac{2}{3}$ feet per second. (2) $58\frac{2}{3}$ feet per second for the first hour, and $14\frac{2}{3}$ feet per second for the second hour.

69. $1624\frac{8}{15}$ feet per second.   **70.** 5796 feet.

71. 83·3 metres.   **72.** 490 metres.   **73.** 10 seconds.

74. (1) 402·5 feet; (2) 20 seconds.   **76.** 321 feet.

77. 19·6 metres towards the ground.

83. Height, 490 metres; velocity, 98 metres per second.

84. 95·1 metres per second.

87. If AB be a line representing in magnitude and direction the current, and AC a line at right angles to the first, representing the direction of the boat's motion; then, from B with radius equal to the velocity of the rowing describe an arc cutting AC at D; BD is the direction in which the boat should be rowed.

89. $$u = w\frac{\sin\beta}{\sin\phi}; \quad v = w\frac{\sin\alpha}{\sin\phi}.$$

90. $w = 86\cdot22 : a = 34°\ 15'\ 36'' : \beta = 19°\ 44'\ 24''$.

91. Each velocity $= 7\cdot07$; each angle $= 45°$.

93. From the southwest, with a velocity of $4\sqrt{2}$ miles per hour.

94. He should aim to a point 17·605 feet in advance of the deer, or along a line making an angle of $1°\ 40'\ 49''$ with the deer's direction.

95. If $u =$ velocity of tube, $v =$ that of the particle; then $\dfrac{u}{v} =$ sine of the angle of inclination.

99. $3600\,g$.   **100.** Times, 2^s and 4^s; velocities, g and $-g$.

101. 87·662 metres.   **102.** $10\sqrt{g}$ feet per second.

103. $1\cdot267^s$.   **104.** $\dfrac{a}{n} + \sqrt{2\,g\,a}$ feet per second.

109. Height $= a$.

ANSWERS AND SOLUTIONS. 117

In exercises on projectiles, let v denote the velocity of projection, a the angle between the direction of v and the horizon.

114. $\dfrac{v^2 \sin^2 a}{2g}$. **115.** $\dfrac{2v \sin a}{g}$.

116. $\dfrac{v^2}{g} \sin 2a$. **118.** 7764 feet.

119. If u and v denote the velocities, and a β the angles of projection respectively; then, distance at time $t = \sqrt{[u^2 + v^2 - 2uv \cos(a - \beta)]}\, t$.

122. $-4°\,54'$. **123.** 233·238 feet.

Lesson IV.

144. 739.24 kilogrammes. **145.** (1) 480; (2) 48·98.
146. 2·3 nearly.
147. 0·45359 m; m;   0·45359 m;   32·19 m; 9·81 × 0·45359 m.
148. 27·734 feet.
149. The rate of change is *mass* × *acceleration* per second.
150. 772 800 000. **151.** 10·2ª; 10·2 times the first force.
157. (1) *g is a number expressing the velocity produced in a falling body in unit of time.*
(2) *g is a number expressing twice the distance through which a body falls in unit of time.*
(3) *g is a number expressing the weight of unit of mass in dynamical measure.*

160. $\dfrac{ma}{g}$ lbs. **161.** $\tfrac{1}{3} \times \tfrac{1}{2} \times g t^2 = 49$; $v = \tfrac{1}{3} g t$.

162. $s = \dfrac{n}{m} \times \dfrac{g t^2}{2}$; $v = \dfrac{n}{m} g t$. **163.** 103·04 feet.

164. 17·96 lbs. **165.** pressure $= m \left(1 + \dfrac{f}{g}\right)$.

166. Time of revolution $= 2\pi \sqrt{\dfrac{Pl}{Qg}}$.

Lesson V.

168. $P \dfrac{f}{a}$. **172.** 7 and 5. **173.** No. **175.** 17.

176. The same and the opposite directions respectively.

178. Draw from any point, A, three lines representing the directions of the two forces and their resultant, and take, on the line which represents the direction of the given force, a length, AB, to represent the magnitude of this force; from B draw a line parallel to the direction of the other force till it intersects the line of direction of the resultant at C; then AC is the magnitude of the resultant.

180. 27 and 120.   **182.** Each force $= \dfrac{1752}{\sqrt{2}}$.

183. Their lines of action must form an equilateral triangle.

185. 80 lbs.   **186.** $5\sqrt{2}$ lbs.

187. Magnitude $= 4\sqrt{2}$; direction is perpendicular to one of the sides of the square.

191. 38 and 114.

206. (1) $P+Q$, (2) $Q-P$, (3) $P-Q$.

208. 4 metres from the end to which the weight of 9 kilogrammes is attached.

209. 4 feet.   **210.** 7 m. from the given pressure.

213. $6\frac{3}{4}$ inches from the end supporting the weight of 5 lbs.

214. $84\frac{1}{8}$ and $88\frac{7}{8}$ lbs.   **216.** 40 lbs.

221. 1792 lbs.   **222.** 255.   **224.** $16\frac{3}{8}$ lbs.

225. 45°.   **229.** $\dfrac{25}{3\pi}$.   **235.** $2P$.

236. 6·89 lbs.; angle of R with first force, 102° 16′.

238. 67·2 lbs.   **239.** $\dfrac{10}{\sqrt{3}}$ kilogrammes.

240. Horizontal thrust $= \frac{1}{2}W \cot a$, where $a =$ inclination of either rafter to the horizon.

241. (1) Pressure $= W \operatorname{cosec} a$, where $W =$ weight of hat, $a =$ semi-angle of the cone.
(2) The pressure becomes infinite.

242. If P, Q, S, are the forces, R their resultant, then, —
$$R^2 = P^2 + Q^2 + S^2.$$

ANSWERS AND SOLUTIONS. 119

243. $R^2 = (\Sigma\ P \cos \alpha)^2 + (\Sigma\ P \cos \beta)^2 + (\Sigma\ P \cos \gamma)^2$

$\cos \phi = \dfrac{\Sigma\ P \cos \alpha}{R},\ \cos \psi = \dfrac{\Sigma\ P \cos \beta}{R},\ \cos \omega = \dfrac{\Sigma\ P \cos \gamma}{R}.$

244. *It is necessary and sufficient for the equilibrium of any number of forces applied to a point, that the sums of the resolved parts of the forces in the directions of three intersecting lines not in one plane be separately equal to zero.*

253. The point of attachment should be as far above the ground as the man is from the foot of the tree.

254. Pressure on 1st plane $= \dfrac{W \sin \beta}{\sin (\alpha + \beta)}.$

Pressure on 2d plane $= \dfrac{W \sin \alpha}{\sin (\alpha + \beta)}.$

255. Let $\theta = $ angle which a line drawn from W to centre of hemisphere makes with the horizon; then, —

$$\cos \tfrac{1}{2} \theta = \dfrac{P \pm \sqrt{(8\ W^2 + P^2}}{4\ W}.$$

256. Let $W = $ weight of each beam, $\theta = $ its angle with the horizon, then, horizontal thrust $= \tfrac{1}{2}\ W \cot \theta.$

257. Let $\theta = $ angle of m with horizon; then, —

$$\tan \theta = \dfrac{m^2 + n^2 \cos \phi}{n^2 \sin \phi}.$$

258. Let $W = $ weight of the beam, $P = $ pressure of beam on the rail, $R = $ pressure of beam on the wall, $\theta = $ angle of beam with the wall, $a = $ half the length of the beam, $b = $ distance of the rail from the wall; then, —

$$\sin \theta = \sqrt{\dfrac{b}{a}};\quad P = W \operatorname{cosec} \theta;\quad R = W \cot \theta.$$

259. Let W denote the weight of the ladder, m and n the segments into which its centre of gravity divides it, θ the angle which the ladder makes with the floor, P the horizontal force required; then, —

$$P = \dfrac{m}{m + n}\ W \cot \theta.$$

ELEMENTARY PHYSICS.

260. (1) Between W and infinity.
(2) " 0 and W.
(3) " 0 and infinity.

261. $\dfrac{P}{W} = \dfrac{\text{product of the weight arms}}{\text{product of the power arms}}$.

262. As 1 to 2.

264. Let P denote the power applied to the wheel, a and b radii of wheel and axle, a the angle of the inclined plane with the horizon, W the weight on the plane; then, —

$$P = \frac{W b \sin a}{a}.$$

267. $\dfrac{W}{40} + \dfrac{1}{20}$ of the pressure on the plane.

268. The weights are as the lengths of the planes.

269. π inches.   **270.** 186·86 lbs.

Lesson VI.

275. $\frac{6003}{32003}$ of a foot from the ship.

376. 2·8 feet per second.

278. By pushing against the air with the palms of his hands.

282. Let m, m', denote the masses, v, v', their velocities, then, —

$$\text{velocity after impact} = \frac{m v + m' v'}{m + m'}.$$

283. 8 metres per second.   **284.** 15 metres per second.

289. Acceleration $= \dfrac{m - m'}{m + m'} g$;   tension $= \dfrac{2 m m'}{m + m'} g$.

290. It will be nothing.

291. Acceleration $= \dfrac{m \sin a - m' \sin a'}{m + m'} g$;

tension of the string $= \dfrac{m m' (\sin a + \sin a')}{m + m'} g$.

ANSWERS AND SOLUTIONS. 121

298. Let u, v, denote the velocities before and after impact respectively, i, ν, the angles of incidence and reflection; then, —
$$\tan i = e \tan \nu,$$
$$v^2 = u^2 (\sin^2 i + e^2 \cos^2 i).$$

299. $\frac{2}{3} b$.

Lesson VII.

304. $90°$. **307.** 5. **308.** The parts must be equal.

321. The centrifugal force at the equator diminishes the weight of a body by about $\frac{1}{289}$th part.

322. 17 times faster than it now revolves.

323. The component of centrifugal force diminishing gravity $= \dfrac{4\pi^2 R}{T^2} \cos 2\lambda$, where $R =$ earth's radius, $T =$ time of revolution, $\lambda =$ latitude.

325. The total variation amounts to $\frac{1}{194}$th part of the weight of a body. In other words, a body which weighs 194 lbs. on the equator would weigh 195 lbs., very nearly, at the pole. Centrifugal force causes a part of this difference, and the variation in the earth's attractive force due to its spheroidal shape produces the remainder of the difference.

Lesson VIII.

328. Let P and Q be the weights of the bodies, T the tension of the string, ϕ the acceleration; then, —

$$T = \frac{2PQ}{P+Q}. \qquad \phi = \frac{P-Q}{P+Q} g.$$

329. 16 oz. **330.** 4 lbs.
332. $\frac{1}{5} g$. **334.** Nothing.

Lesson IX.

337. At the centre of figure in each case.
339. At a point 8 inches from the 3 lbs.

ELEMENTARY PHYSICS.

349. 3600 times the length of the second's pendulum.
350. Fourfold.   **351.** 9·8098.
353. At a distance of 21 inches from the 40 lbs.
354. $2\,b$.

355. $\dfrac{b\,l}{a+b}$, where $l =$ length of rod, $a =$ its weight, $b =$ weight of beetle.

356. Let $D =$ density of the sphere whose radius is 8 inches, $D' =$ density of the other sphere; then their common centre of gravity is at a distance of $\dfrac{540\,D'}{8\,D + 27\,D'}$ from the 8-inch sphere.

357. $2\,r\,\text{cosec}\,\theta$.   **359.** Weight lost $= W\,\dfrac{(a-b)^2}{a\,b}$.

361. As 144 to $145\tfrac{1}{5\,1\,6}$.   **362.** As $m^2\,l$ to $m'^2\,l'$.

364. $\dfrac{9\,h}{2200}$, h being expressed in feet.

365. 0·94 of a mile.

367. Earth's radius $= \dfrac{h\,t_2^2}{t_2^2 - t_1^2}$, where t_1 is the time of an oscillation at the surface, and t_2 at the given depth h.

Lesson X.

372. 1.   **373.** 18 lbs.; 0·75.
374. Twice as strong.

Lesson XI.

383. 25 kilogrammes.   **384.** 1 oz.
390. 3 kilogrammes per sq. centimetre.   **391.** 100 metres.
392. 6400 kilogrammes.   **393.** 7500 grammes.
395. As 8 : 1, in both cases.   **397.** 348160 π grammes.
403. 10000 kilogrammes.   **406.** 18 grms.   **407.** 725 grms.
408. 0·8 gramme.   **409.** 2·8.
410. The smaller, in mercury; and the larger, in a vacuum.

411. 9868 cubic centimetres.   **412.** $\dfrac{W}{\pi\,r^2}$.

ANSWERS AND SOLUTIONS. 123

413. $166\frac{2}{3}$ kilogrammes.

414. Let V_1, V_2, V_3, &c., be the given volumes, σ_1, σ_2, σ_3, &c., the specific gravities, σ' the specific gravity of the mixture; then, —

$$\sigma' = \frac{V_1 \sigma_1 + V_2 \sigma_2 + V_3 \sigma_3 + \&c.}{V_1 + V_2 + V_3 + \&c.}.$$

415. Let W_1, W_2, W_3, &c., be the given weights, σ_1, σ_2, σ_3, &c., the specific gravities, σ' the specific gravity of the mixture; then, —

$$\sigma' = \frac{W_1 + W_2 + W_3 + \&c.}{\dfrac{W_1}{\sigma_1} + \dfrac{W_2}{\sigma_2} + \dfrac{W_3}{\sigma_3} + \&c.}$$

416. 0·8975.   **417.** 0·8857.   **418.** 1·2.
419. 7511 grammes of gold, and 2489 grammes of silver.
420. 3·5 grammes.   **421.** 11·2.   **423.** $\frac{8}{5}$.
424. 1·9 nearly.   **425.** 0.792.

428. The dividing line must be at the depth $\dfrac{h}{\sqrt{2}}$, where h denotes the vertical edge.

429. Let h = the vertical edge; then the lines of division are at the depths

$$\sqrt{\frac{1}{n}}\,h, \qquad \sqrt{\frac{2}{n}}\,h, \qquad \sqrt{\frac{3}{n}}\,h, \&c.$$

431. $\dfrac{\text{Total pressure}}{\text{Weight of the liquid}} = \dfrac{h+r}{r}.$

432. $\dfrac{\text{Total pressure}}{\text{Weight of the liquid}} = \dfrac{2\sqrt{r^2+h^2}+r}{r}.$

437. 947076 tonnes.
438. Zinc 17·82 grammes, copper 32·18 grammes.
439. If W, W' and σ, σ' denote the weights and specific gravities respectively of the two bodies, then the required specific gravities are, —

(1) $\dfrac{m}{m-1} \cdot \dfrac{W+W'}{\dfrac{W}{\sigma}+\dfrac{W'}{\sigma'}}$ (2) $\dfrac{n}{n+1} \cdot \dfrac{W+W'}{\dfrac{W}{\sigma}+\dfrac{W'}{\sigma'}}.$

124 ELEMENTARY PHYSICS.

440. 1·51 nearly. **441.** $P \cdot \dfrac{Qr - Rq}{Qp - Pq}$. **442.** $\dfrac{W' - W}{\sigma - \rho}$.

443. If σ' = specific gravity of the material of which the weights are made, ρ = specific gravity of the air; then, —

$$\text{true weight} = P \cdot \frac{\sigma}{\sigma'} \cdot \frac{\sigma' - \rho}{\sigma - \rho}.$$

444. 11·85 centimetres.   **445.** 16·1 square centimetres.
446. (1) 0·9 of the height; (2) 0·647 of the height.
447. The altitudes of the two cones are inversely as the cube roots of the specific gravities of the cone and the liquid, whatever be the vertex-angle of the cone.
448. Thickness : diameter = $\tfrac{1}{2}(1 - \sqrt[3]{\sigma - 1}) : 1$.
449. 1·252 seconds.

Lesson XII.

453. 103·33 kilogrammes.   **454.** 14684 kilogrammes.
455. 13·596 gh.   **456.** 1033·3 centimetres.
457. 41·51 feet.   **458.** 245 centimetres.
460. Diminished.   **461.** Neither.
463. 187 feet 9 inches.   **465.** See Appendix V.

466. $\dfrac{2000\,\pi}{3}$ grammes, nearly.

468. If we have a volume V of gas at the pressure h, and then change the volume to V' the pressure becoming h'; then Mariotte's law asserts that, —

$$V : V' = h' : h, \quad \text{or} \quad \frac{Vh}{V'h'} = 1.$$

But experiments have shown that gases, except hydrogen, are more compressible than the law indicates, that is, for all gases except hydrogen $\dfrac{Vh}{V'h'} < 1$.

But for hydrogen $\dfrac{Vh}{V'h'} > 1$.

470. 1728 times the initial pressure.
471. 6·354 litres.   **472.** The air will flow out.

ANSWERS AND SOLUTIONS. 125

473. More of it will be submerged.
474. The tension will be diminished.
476. Let h = depth descended in feet, k = height of a water barometer, in feet; then, —

$$\frac{\text{height ascended in the bell}}{\text{total height of the bell}} = \frac{h}{h+k}.$$

477. 46·968 centimetres.
478. Let V = volume of the receiver, v = volume of the barrel, D = initial density of the air in the receiver, D_n = the density of the air after n strokes; then, —

$$D_n = D \left(\frac{V}{V+v}\right)^n.$$

481. $6\frac{21}{64}$ lbs. per sq. inch. **482.** As 5874 to 1.
483. It is equal to the weight of a column of water having a section equal to the area of the piston, and a height equal to the height of the column of water raised.
485. Not more than 76 centimetres, by the suction principle.
486. The siphon would cease working and the water in the arms would flow out.
487. The rapidity of the flow, and the possible height of the short arm, would be diminished.
488. The siphon would work backwards, that is, the flow would be from the long arm into, and out of, the short arm.
489. 786·875 kilogrammes.
494. If d = diameter at depth a, d' = diameter at depth $\frac{a}{4}$, h = height of a water barometer; then, —

$$d' = d \sqrt[3]{\frac{h+a}{h+\frac{a}{4}}}.$$

496. Let l = length of the tube, h = height of a perfect barometer, y = length of the air-column in the tube at the pressure h, x = depression produced by the air; then, —

$$x = \tfrac{1}{2}(h+l) \pm \tfrac{1}{2}\sqrt{(h+l)^2 - 4h(l-y)}.$$

497. $\quad a + \dfrac{h - h''}{h - h'}(a' - a).$

498. $\dfrac{\text{Initial density of the air}}{\text{Final density attainable}} = \dfrac{\text{capacity of barrel}}{\text{untraversed space}}.$

499. Let V, v, denote the volumes of the receiver and barrel respectively, v' the untraversed space, D the initial density of the air, D_n the density after n strokes; then,—

$$D_n = D\left[\dfrac{V^n}{(V+v)^n} + \dfrac{v'}{v}\left(1 - \dfrac{V^n}{(V+v)^n}\right)\right].$$

500. Let A and B be the capacities of the receiver and barrel respectively, h the measure of the pressure of the atmosphere; then,—

$$\text{tension after } n \text{ strokes} = \dfrac{A + nB}{A} \cdot h.$$

501. 350 strokes.

502. The general condition is, that the pump will not work unless the play of the piston be greater than the square of the distance from the surface of the water in the well to the highest position of the piston divided by four times the height of the water-barometer.

II. — SOLUTIONS.

Lesson I.

19. The unit of mass in the Metric System (the kilogramme), which is, strictly speaking, the quantity of matter in a certain platinum weight kept in Paris, was intended to be, and may be taken as, equal to the mass of one unit of volume (the litre) of pure water at 4° C.

Hence, 2 litres of water weigh 2 kilogrammes, and in general a litres weigh a kilogrammes.

21. The density of a body is the *ratio of its mass to its volume*, or, in symbols,

$$D = \frac{M}{V},$$

D denoting density, M mass, and V volume.

The numerical measure of the density of a substance is obtained by taking the unit of volume, or putting $V = 1$, in which case $D = M$, or the measure of density is the *number of units of mass in unit of volume of the substance*.

The value of the density of a substance evidently depends on the units of mass and volume adopted, but density is always the ratio of a mass to the cube of a length, or, as it has been expressed, the *dimensions of density* are $\frac{M}{L^3}$, M denoting a mass, and L a length.

23. Because the unit of mass in the Metric System is the mass of unit of volume of pure water at 4° C.; or, in other words, because, in the case of water, when $V = 1$, $M = 1$, and therefore, —

$$D = \frac{M}{V} = \frac{1}{1} = 1.$$

26. The word *weight*, both in scientific and common language, is usually employed to denote *mass* or quantity of matter. But the weight of a body, properly speaking, is the *measure of the force with which the body is drawn towards the centre of the earth*. This force is different in different places, being greater, for example, at the pole than at the equator, and greater at the level of the sea than at the top of a mountain. The *mass* of a body is the *quantity of matter which it contains;* this is invariable, and would remain the same even if the force of gravity did not exist, in which case bodies would have no weight.

Lesson III.

Representation of Velocities and Forces. — A velocity (or a force) is said to be represented geometrically by drawing a straight line in the direction of the given velocity (or in the direction in which the given force acts) and making the line as many units long as there are units in the given velocity (or in the given force). For example, a velocity of 10 feet per second towards the north would be represented on paper by a line 10 inches long, drawn as in maps, &c., perpendicular to the upper edge of the paper. A force of 16 units acting toward the east might be represented on paper of moderate size by a line 4 inches long, taking for convenience one quarter of an inch to represent the unit of force, the line being drawn from a point where the force is supposed to act, towards the right-hand edge of the paper and perpendicular to that edge. An arrow-head may be added, as in Fig. 2, to show which of the two directions of the line is to be taken. When forces represented by lines lettered at their extremities are referred to, the order of the letters indicates the direction of action. Thus, a force AB means a force acting *from A towards B.*

56. The Parallelogram of Velocities may be stated as follows : —

If two velocities be represented in magnitude and direction by two straight lines drawn from any point, the diagonal of the parallelogram constructed upon these two lines will represent the resultant velocity in magnitude and direction.

This proposition is one of the utmost importance, inasmuch as the great majority of mechanical problems, whether of the practical kind which require solution in the operations of Civil Engineering, or of a purely theoretical nature, involve the action of several forces in different directions.

Mr. Stewart has given, in Lesson IV., § 24, a partial proof for the case of two simultaneous continuously acting forces.

ANSWERS AND SOLUTIONS. 129

The proof there given is imperfect, because it is only shown that the body will be at the extremity of the diagonal at the end of the time considered, not that it will constantly be on this diagonal throughout the interval. This is a very natural inference, it is true, but it is an inference which can be deduced legitimately from principles more general, if not more axiomatic in their character.

In the example of composition of velocities selected by Mr. Stewart, the velocities are generated by gravity and by magnetic attraction, — that is, by continuously acting forces, — and consequently the velocities generated are not uniform but accelerated. In other words, Mr. Stewart's proposition is not the Parallelogram of Velocities but of Accelerations, — a proposition equally true, but usually and properly considered a logical deduction from the more simple case of uniform motion.

Sir Isaac Newton, taking the case of uniform velocities, has given in his "Principia" the following simple and convincing proof of this famous proposition. It will be noticed that he obtains the result as a deduction from his First and Second Laws of Motion.

Suppose that a body, in a given time, by the effect of a single force M impressed at A, would move with a uniform velocity from A to B; and suppose that the body, in the same time, by the effect of another single force N impressed at A would move with a uniform velocity from A to C; then if both forces act simultaneously at A, the body will move uniformly in the given time along the diagonal from A to D.

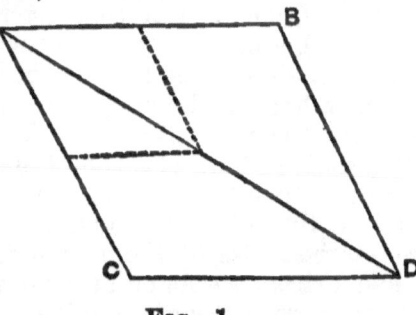

FIG. 1.

For since the force N acts in the direction of the line A C parallel to B D, this force, by the Second Law, will not at all alter the velocity generated by the other force M by which the body is carried towards the line B D. The body, therefore, will arrive at the line B D in the same time whether the force N be impressed or not; and therefore at the end of that time it will be found somewhere in the line B D. By the same reasoning, at the end of the same time it will be found somewhere in the line C D. Therefore it will be found at the point D where both lines meet. And it will move in a straight line from A to D by the First Law.

It will be observed that Newton supposes that the two forces act *instantaneously* ; that is, are of the nature of *blows*, as that of a bat upon a ball. Such a force communicates its effect in a time too small to be taken account of, and is in this respect totally different from a force like gravity or attractions of any kind which act *continuously*. Forces of the former kind are often called *impulsive* forces, and forces which act during finite periods of time, like gravity, are called *continuous* forces. Impulsive forces tend to produce uniform motion, being only prevented from so doing by the universal presence of retarding causes, such as friction, resistance of the air, &c. Continuous forces tend to produce accelerated motions. In Newton's exposition, the uniform motions are supposed to be produced in the simplest possible way, by the action of impulsive forces, combined with the absence of retarding causes. But in nature no such instances can occur, because retarding causes always exist. Bodies may move with uniform velocities, but the uniform motion must arise, in every case, from the fact that several forces are acting on the body in such a manner that they neutralize each other's effects and leave the body free to obey the First Law of Motion. For instance, a horse is drawing a load along a road at a uniform rate ; here the muscular effort of the horse is just balanced by the friction of the wheels on the ground, and the resistance of the air, which latter is, of course, very trifling.

In fact, the reference which Newton makes in his proof to force as the cause of the motion is unnecessary ; and, inasmuch as it has been found of great advantage in Mechanics to treat many properties of "motion, displacement, and deformation" independently of force, mass, &c., under the head of Kinematics, or the Geometry of Motion, we will present the solution of the Parallelogram of Velocities free from any reference to force.

Suppose that a body at A (Fig. 1) moves with uniform velocity from A to B, and that simultaneously the line A B moves uniformly, and parallel to itself, in the direction A C or B D. Suppose also that in the time required for the body to move from

A to B, the line would move from the position A B to the position C D. Then if the line remained at rest, the body would be at B at the end of the time considered, but since by the motion of the line the point B takes the position D, and since both motions take place independently of one another, the body will be found at the end of the interval at the point D. This is true *whatever be the time considered*, provided both motions are uniform. If, then, we take one half the interval of time already considered, the velocities represented by A B and A C will each be reduced one half, and by drawing the dotted lines we obtain a parallelogram similar to the first one by Geometry. Therefore its diagonal is equal to $\frac{1}{2}$ A D ; that is, the time being halved, the space passed over by the body is halved, the direction of the body's motion being unchanged. The same reasoning may be applied to any and every fractional part of the time considered ; therefore the body will move uniformly in a straight line from A to D.

58. Let the velocities be represented by the lines O A, O B, O C, and O D in Fig. 2. (The line O H is not employed in this Exercise.)

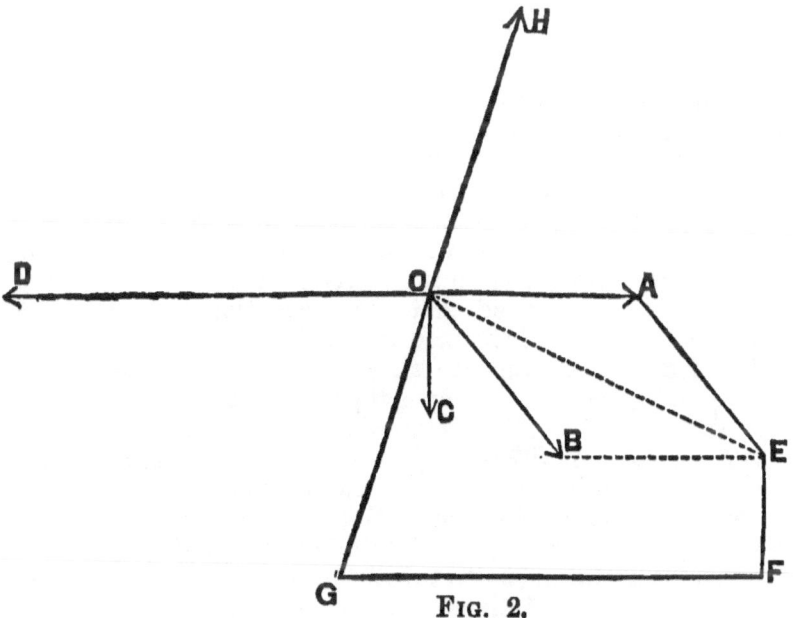

Fig. 2.

Find, by the Parallelogram of Velocities, the resultant O E of the first two velocities O A and O B. Then compound by the same principle this resultant with the velocity O C, obtaining O F as the resultant of the first three velocities. Proceeding in the same way with the remaining velocity O D, we finally obtain O G as the resultant or single velocity equivalent to all the simultaneous velocities. This method is evidently applicable to any number of simultaneous velocities impressed upon a body at O.

It should be observed, that, speaking strictly, it is impossible for a body to be at, or in, a point O. Bodies are finite portions of matter and have finite magnitudes. And, in point of fact, if the several velocities above represented were simultaneously impressed on a body in directions having a common point of intersection O within the body, O N would *not* represent the actual resultant motion of the body (unless O were the centre of gravity of the body), on account of the connections of the parts of the body due to cohesion, and the consequent mutual actions of those parts. The determination of the actual motion of the body is a problem of far greater difficulty than the simple operation of finding a resultant by the Parallelogram of Velocities. But this principle is a first and an essential step in the chain of reasoning which leads to the solution of any case of motion however complex. The first thing to be done, therefore, is to set forth the fundamental principle in as simple a form as possible. Most writers on Mechanics do this by the aid of the conception of *material points* or *particles*, which are defined as bodies so small that their dimensions may be neglected. Bodies are regarded as composed of an indefinitely large number of particles. It will be noticed that the conception of a particle is different from that of the molecule in Chemistry.

If, then, we use language strictly, instead of speaking of a "body at O," we should say a "particle at O," or "a material point at O," or, for shortness' sake, "a point at O."

59. In Exercise **56** the object was to find a single velocity which was equivalent to two simultaneous velocities; this is frequently called the Composition of Velocities. It is often necessary for the purposes of proof or illustration to perform the reverse operation, — to substitute for a single velocity two velocities in assigned directions.

For instance, a ship is sailing at the rate of 8 miles per hour in a direction 30° to the east of south; at what rate is she moving towards the east and towards the south respectively? Questions of this kind fall under the general problem of the Resolution of Velocities; this problem is solved geometrically as follows:—

Let O C represent the velocity which we wish to resolve in the directions O X and O Y; from C draw the lines C A and C B parallel to O Y and O X respectively. Then, by the Parallelogram of Velocities, O C is the resultant of velocities represented by O A and O B, so that O A and O B are the resolved parts required. Resolution in directions making a right angle with each other is by far the most common and useful, being attended with this great and obvious advantage, that the components O A and O B are wholly independent of each other, so that each component represents the entire effect of the velocity O C, estimated in its own direction. In the question above asked, it can be easily found that the ship is moving eastward at the rate of 4 miles an hour and at the same time moving southward at the rate of $4\sqrt{3}$ miles an hour.

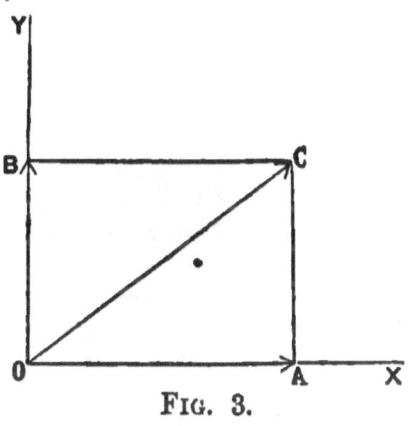

FIG. 3.

60. The resultant of the velocities O A and O B, in Fig. 4, is a velocity represented by the third side O D of the triangle O A D. This resultant would be neutralized by an equal and opposite velocity O C impressed upon the particle at O, so that O A, O B, and O C form a system of three velocities such that, if impressed simultaneously upon the particle at O, the particle would remain at rest. Now taking the sides of the triangle O A D in order, O A represents the first velocity, A D represents the second, because it is parallel and equal to O B, and D O represents the third velocity O C, which is by supposition equal to D O and has the same direction.

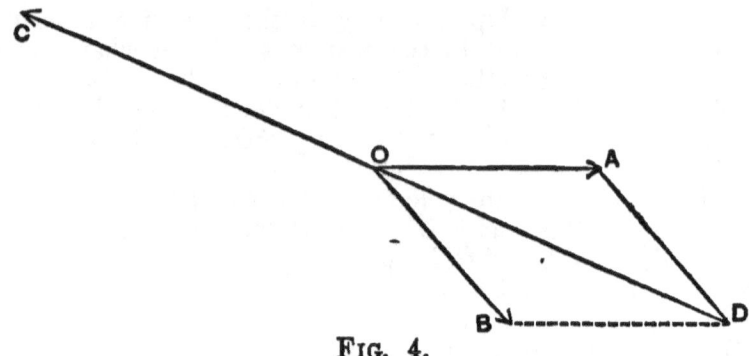

FIG. 4.

61. This is merely an extension of the process of proof in Exercise **60.** Use Fig. 2 for illustration.

62. The term *projection* is one of great importance in certain branches of study, especially Descriptive Geometry and Mechanical Drawing. What we are here concerned with is the projection of one line on another; this may be defined by reference to Fig. 5. Let A B be a line of given length. To find its projection on any line, X Y, let fall from A and B lines perpendicular to X Y and meeting X Y at the points M and N. Then M N is the projection of A B on X Y.

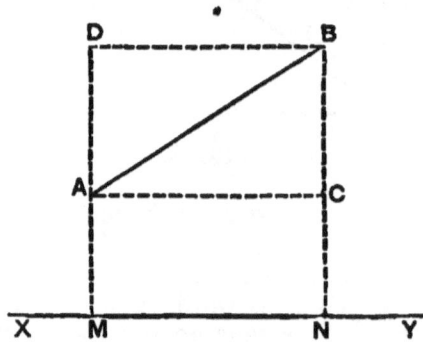

FIG. 5.

Now it is clear (as was stated in the Solution of Exercise **59**) that the effective component of a velocity in any direction will be found by resolving the velocity into two components; one in the given direction, which will be the effective component sought, the other in a direction perpendicular to the given direction, which component is wholly independent of the former. If we resolve A B into these two components by the aid of the Parallelogram of Velocities, we find them to be A C and A D, of which the former, A C, is the

ANSWERS AND SOLUTIONS. 135

effective component in the direction X Y, and is equal to M N, or the projection of A B on X Y.

The examination of the special case referred to is left to the student.

67. A velocity may change, (1) in quantity, (2) in direction. We are here concerned only with the first kind of change, the motion being supposed to be rectilinear. For curvilinear motion, see Exercises **124 – 129**. Acceleration is the *rate of change of velocity*. Velocity may change in quantity either by increasing or by diminishing, but the term "acceleration" may be extended to include both cases by applying to velocities and accelerations the algebraic convention of positive and negative signs to denote opposite directions. Acceleration of velocity may be either uniform or variable : it is said to be *uniform* when the point receives equal increments of velocity in equal times, and is then measured by the *increase of velocity per unit of time*. To illustrate, take the force of gravity, and let the direction of this force (towards the earth's centre) be considered positive. It is known that a body, let fall from a point above the earth's surface, acquires a velocity of 32 feet per second in the first second; that is, at the end of the first second it is moving at such a rate that it would move over 32 feet during the next second, if its motion were uniform. But its motion is not uniform ; gravity exerts the same effect upon it during the second as during the first second, so that at the end of the second second its velocity is 64 feet per second ; and at the end of the third second it would be 96 feet per second, and so on. In this case the acceleration, or change of velocity per unit of time, is uniform, and is +32 feet per second in each second. Which direction shall be considered positive in any case is purely arbitrary, and is usually selected on grounds of convenience. In the example just given, where gravity was the only force concerned, and the motion produced by gravity the only motion, the simple and obvious course was to call the direction of gravity the positive direction. But take the familiar case of a stone thrown vertically upwards. This is an example of retarded motion. But the analysis of the case shows that there are two motions, simultaneous, independent of each other, and in opposite directions, — one a uniform motion vertically upwards due to the muscular impulse of the hand, the other a uniformly accelerated motion vertically downwards due to the constant action of the force of gravity. We may take either direction as positive; then the opposite direction must be considered negative. It is rather more convenient, probably, to call the upward direction positive; and it is certainly more natural to consider the direction of the actual motion as positive, and retarded motion as an instance of negative acceleration. Let this, then, be assumed, and suppose a stone thrown vertically upwards with a velocity of 96 feet per second. It has two simultaneous motions, one represented by the uniform velocity +96, the other represented by the uniform acceleration −32. It is evident that at the end of 3 seconds the stone will have a velocity of −96 downwards, produced by gravity, which will just cancel the velocity +96, that is, the stone will come to rest. The space passed over by the stone is quite another matter, and is investigated in Exercises **78** and **80**.

68. The *average* acceleration during any time is the total velocity gained during that time divided by the time.

71. In 9 seconds a body will fall through $\frac{1}{2}$ (9·8) 81 metres; in 8 seconds it will fall through $\frac{1}{2}$ (9·8) 64 metres. Therefore, during the *ninth* second it falls through $\frac{1}{2}$ (9·8) (81 − 64) metres = 4·9 × 17 metres = 83·3 metres.

75. Here, and in **78** and **79**, the notation is employed with special references to the most important and universal of all forces, that of gravity. The initial velocity a is supposed to be in the direction of gravity. The letter g is universally employed by writers on physical science, as in this Exercise, to denote the acceleration of gravity, which varies slightly with the locality, and is only expressible approximately by decimals not needed in most investigations. g, then, denotes the number of units of velocity generated by the force of gravity in one second. The force being constant, the number generated each second will be the same, and therefore at the end of t seconds the velocity acquired by the body will be $g t$ units. This is the entire effect of the force of gravity, and this effect is by the Second Law of Motion the same whether the body be initially at rest or moving in any direction with any velocity. In this case the body is supposed to have initially a uniform velocity of a units, which velocity it would continue to have forever, by First Law, were no forces to act upon it. It is evident, then, that as the initial and acquired

ANSWERS AND SOLUTIONS. 137

velocities are in the same direction, the resultant velocity at the end of t seconds will be their sum, or $a + gt$. This formula includes the case in which the initial velocity a is opposite to the direction of g simply by considering either a or g negative, as pointed out in the Solution of Exercise **67**.

78. We may employ here the principle of the Independence of Motions (or the Effects of Forces), stated by Mr. Stewart at the beginning of Lesson III., and there called by him the Second Law of Motion, although the Law as given by Newton contains, or at least implies, much more. (See Solutions of **135** and **138**.) Where several forces act on a body, estimate separately the effect of each force in producing motion; then combine these effects by addition or subtraction, if the forces act along one line, or by the aid of the Parallelogram of Velocities, if they act along different lines. In this case we have two motions along the same line. The point at the instant when the time t begins to be reckoned is moving with a uniform velocity a due to a force (the impulse of the hand, say) which has ceased to act. From this instant the effect of gravity is also to be taken into account. In virtue of the uniform velocity a the point will describe the space at in the time t. In virtue of the action of gravity, it will describe the space $\frac{1}{2}gt^2$ in the time t, as shown by Mr. Stewart in Lesson III. Therefore the total space described is $at + \frac{1}{2}gt^2$.

We may prove this theorem without reference to the Laws of Motion, as follows: the *average* velocity of a point during any interval of time is the space described during that time divided by the time. In this case, since the acceleration is uniform, the average velocity is the arithmetical mean between the initial and final velocities. For as the velocity increases uniformly, its value at any time *before* the middle of the interval is as much *less* than this mean as its value at the same time *after* the middle of the interval is *greater* than the mean. Here the initial velocity is a, and the final velocity $a + gt$. Therefore the average velocity $= \frac{1}{2}(a + a + gt) = a + \frac{1}{2}gt$. And space described = average velocity $\times$ time $= at + \frac{1}{2}gt^2$.

80. Here the initial velocity is opposite to the direction of gravity. Consider the direction of projection positive, and that of gravity negative. Then the general formulae of **75** and **78** must be written, —

$$(1) \quad v = a - gt$$
$$(2) \quad s = at - \tfrac{1}{2}gt^2$$

The time of ascent is evidently the value of t in (1) when $v = 0$, or $\frac{a}{g}$. The height ascended is found by substituting this value of t in (2).

81. Put $s = 0$ in equation (2) in the preceding solution, and solve with respect to t. We find $t = 0$, or $t = \frac{2a}{g}$. The first value corresponds to the instant of leaving the ground, the second to the instant of reaching it again. But we have already seen that the time of ascent $= \frac{a}{g}$. Therefore time of descent also $= \frac{a}{g}$.

82. If $a =$ velocity of projection, the time of ascent $= \frac{a}{g} =$ time of descent, by **80** and **81**. But if a body falls freely during the time $\frac{a}{g}$, the velocity acquired $= g \times \frac{a}{g} = a$.

Lesson IV.

135. We will here give Newton's Laws of Motion, both in the original Latin of Newton, and as translated by Thomson and Tait in their Treatise on Natural Philosophy. Though called by Newton Laws of Motion, it would be more accurate to call them Laws relating to the Connection of Force with Motion.

Lex I. *Corpus omne perseverare in statu suo quiescendi vel movendi uniformiter in directum, nisi quatenus illud à viribus impressis cogitur statum suum mutare.*

Law I. *Every body continues in its state of rest, or of uniform motion in a straight line, except in so far as it may be compelled by impressed forces to change that state.*

Lex II. *Mutationem motûs proportionalem esse vi motrici impressae et fieri secundum lineam rectam quâ vis illa imprimitur.*

Law II. *Change of motion is proportional to the impressed force, and takes place in the direction of the straight line in which the force acts.*

ANSWERS AND SOLUTIONS. 139

Lex III. *Actioni contrariam semper et aequalem esse reactionem: sive corporum duorum actiones in se mutuò semper esse aequales et in partes contrarias dirigi.*

Law III. *To every action there is always an equal and contrary reaction; or, the mutual actions of any two bodies are always equal and oppositely directed.*

136. Change of motion is determined in the first place by the mode in which the quantity of motion of a moving body is measured. This measure Newton himself explicitly defines in the second of his *Definitions* which precede his *Axiomata sive Leges Motûs*. The *quantity of motion* or *momentum* of a rigid body moving without rotation is, according to Newton, proportional to its mass and velocity conjointly. Thus with a double mass and equal velocity the quantity of motion is double; if the velocity be also doubled, it is quadruple. Take, as unit of momentum, that of unit of mass moving with unit of velocity; then the momentum of m units of mass moving with v units of velocity will be mv, a result agreeing with the definition of momentum given by Mr. Stewart, § 23. By change of motion, then, Newton meant change of momentum, and this change may arise either from change of mass or change of velocity. Mass is an element which is not considered in the motion of material points; in all other cases variations in mass simply produce proportional variations in momenta. What is meant by change of velocity is plain enough so long as the change takes place along the line of motion; the change is to be added to the existing velocity if it takes place in the same direction, subtracted from this velocity if it takes place in the opposite direction. But suppose a velocity change in direction as well as amount; how is this change to be measured? We cannot discuss this question here, but Fig. 1, page 129, will serve to show the character of the answer. If the velocity A B be changed to A D, the change is represented in magnitude and direction by A C. In the example mentioned in the Solution of **59**, if a ship be sailing due east at the rate of 4 miles per hour, and its course be suddenly changed so that it begins to move with a velocity of 8 miles per hour in a direction 60° south of east, the change has been $4\sqrt{3}$ miles per hour due south. This represents what has been added, so to speak, to the easterly motion to produce a velocity of 8 miles per hour in a direction 60° south of east.

ELEMENTARY PHYSICS.

137. This principle is given by Mr. Stewart at the beginning of Lesson III. It is stated yet more generally by Thomson and Tait as follows: "*When any forces whatever act on a body, then, whether the body be originally at rest, or moving with any velocity and in any direction, each force produces in the body the exact change of motion which it would have produced if it had acted singly on the body originally at rest.*" The student should try to find illustrative examples of the truth of this principle additional to those given by the author.

138. The Second Law informs us that a force is proportional to the change of motion which it produces, and change of motion has been explained to be change of momentum in the Solution of **136.** Force, then, is to be measured by the change of momentum which it produces. But as a force produces a continuous change of momentum, we can only compare forces with each other by comparing the changes of momentum produced in some one common interval of time. The simplest interval, and the one universally adopted, is the unit of time, or one second. Thus it appears that the momentum generated in unit of time is the measure of a force, and this is equal to the product of the mass of the moving body and the acceleration, because the change of velocity per unit of time is by definition acceleration.

139. The general definition is *a force which acting for a unit of time upon a unit of mass will generate a unit of velocity.*

141. The statical unit of force (pound or kilogramme) is the weight of unit of mass, or the pressure which unit of mass exerts in consequence of the earth's attraction upon it. Now a unit of mass falling freely acquires a velocity of g units in one second. Therefore the dynamical measure of the force of gravity upon a unit of mass is equal to $1 \times g$ or g units of force. In other words, the *same* force which is expressed in statical measure by 1 unit is expressed in dynamical measure by g units, or

$$g \text{ dynamical units} = 1 \text{ statical unit.}$$

Therefore, $$1 \text{ dynamical unit} = \frac{1 \text{ statical unit}}{g}.$$

The derivation of the rules referred to is left as an exercise for the student.

ANSWERS AND SOLUTIONS. 141

142. The first and fundamental use of the words "pound" and "kilogramme" is to denote *units of mass* or quantities of matter equal to that contained in certain precisely defined standards preserved in the Public Archives with the greatest care. The secondary use is as *units of force* or rather *of pressure*. We are all brought into daily contact with the force of gravity as producing pressure if not motion, and Engineers, Architects, and many other practical men are constantly called upon to measure and compare pressures produced by gravity. It was very natural and very convenient, therefore, to adopt as a unit of force the pressure produced by the standard of mass, or "standard weight," as it is called, and to employ this unit in measuring all kinds of pressures, whether produced by gravity or by the action of other forces. Thus, a force of 20 lbs. is a force just capable of sustaining against gravity a 20-lb. weight. "In all countries," says Prof. Maxwell, "the first measurements of force were made in this way, and a force was described as a force of so many pounds' weight or grammes' weight. It was only after the measurements of forces made by persons in different parts of the world had to be compared, that it was found that the weight of a pound or a gramme is different in different places, and depends on the intensity of gravitation, or the attraction of the earth; so that for purposes of accurate comparison all forces must be reduced to dynamical measure." When great accuracy is required in expressing a force in statical or gravitation measure, it is necessary to specify the locality where the observation is made; thus, so many London pounds of force, so many Paris kilogrammes of force.

143. The *spring balance* is an instrument for measuring *force;* for example, it measures directly the force of gravity upon a body; if the same body be weighed by a spring balance in different latitudes, it will have different weights, because the spring of the balance will be stretched more or less by the changes in the force of gravity. The *common balance* is an instrument for comparing quantities of matter or *masses*. However much the earth's attraction upon a given body may vary, it will vary in the same ratio upon the standard weights employed. If, then, we put a body in one pan of a balance, and equipoise it by placing standard weights in the other pan, the equilibrium will be maintained in all parts of the earth, and everywhere the mass of the body will appear to be, as it really is, invariable.

150. Apply the formula given in Appendix V., at the end of Table II., remembering that in this formula force is supposed to be expressed in dynamical measure, whereas in the Exercise statical measure is used. This formula applies to every case of momentum generated by a constant force like gravity, and is true, since velocity = acceleration × time. The use of this formula for the Exercise shows at once that the mass of the train is not required in order to find the momentum.

153. Since the forces act on a point, mass is eliminated, and, therefore, the forces must be as their accelerations respectively. Moreover, they have the same directions as these accelerations, the direction of a force being defined by the direction of the motion which it tends to produce. The lines representing the accelerations may then be considered to represent the forces also, and hence a single force measured by the resultant acceleration, and in its direction, will be the equivalent of any number of simultaneously acting forces. Now it can be easily shown that the propositions known as the Parallelogram of Velocities and the Polygon of Velocities hold equally true of accelerations. Indeed, this may be at once inferred from the fact that acceleration is merely a change of component velocity in a given direction; hence it is clear that its laws of composition and resolution must be the same as those of velocity. Therefore the laws of the composition and resolution of forces are the same.

Lesson V.

Definitions of Resultant and Balancing Force. — We will give here general definitions of these important terms, which are often mistaken for each other.

I. *Resultant.* — When any number of forces act upon a body, and are not in equilibrium, and when there is one force capable of producing the same effect as the system of forces, this one force is called the *resultant* of the system.

II. *Balancing Force.* — When any number of forces acting upon a body are not in equilibrium, but are capable of being reduced to equilibrium by the application of a single force, this force is called their balancing force.

The resultant and the balancing force of a system evidently form a pair of equal and opposite forces.

ANSWERS AND SOLUTIONS. 143

167. The straight line should be drawn, from the point of application of the force, in the direction of the force, and containing as many units of length as there are units of force.

169. The principles here referred to are generally regarded as axiomatic when the terms in which they are expressed are distinctly understood. We shall so regard them, and shall now give them. They should be carefully learned, on account of their subsequent applications.

I. *Definition of Equal Forces.* *Two forces are said to be equal if, when applied to the same point in opposite directions, they balance one another, or are in equilibrium.* Forces in equilibrium are often called *pressures*.

This definition is only a modified form of the following more general principle, which is Newton's Third Law of Motion applied to forces in equilibrium.

II. *Action and reaction are always equal and opposite.* A table presses against a book just as much as the book against the table; if this were not so, the book and table would move either downwards or upwards, according as the pressure of the book or of the table preponderated.

III. *If a material point or rigid body be acted on by a system of forces, then the additional application of a system of forces in equilibrium will have no effect.*

IV. *A force may be transmitted to any point in the line of its action, without altering its effect on a rigid body.*

V. *The tension of a perfectly flexible cord in contact only with perfectly smooth surfaces is the same throughout its length.*

174. See **153**, and Solution.

188. Two parallel forces are said to be *like* when they act in the same direction, *unlike* when they act in opposite directions.

189. Let P and Q be the forces acting at A and B respectively, A and B being regarded as rigidly connected.

The effect of the forces will not be altered if we apply two equal forces, S, S, at A and B acting along A B in opposite directions (Axiom III.). Compound them with P and Q respectively, and we obtain for the resultants X and Y. Produce the lines of action of these resultants till they meet at D, and draw D C parallel to the lines A P and B Q, meeting

A B at C. Transfer the resultants X and Y to D (Axiom IV.), resolve them along D C and a straight line through C parallel to A B; each of the latter components will be equal to S, and they will act in opposite directions, and will balance each other; the sum of the former components will be $P + Q$.

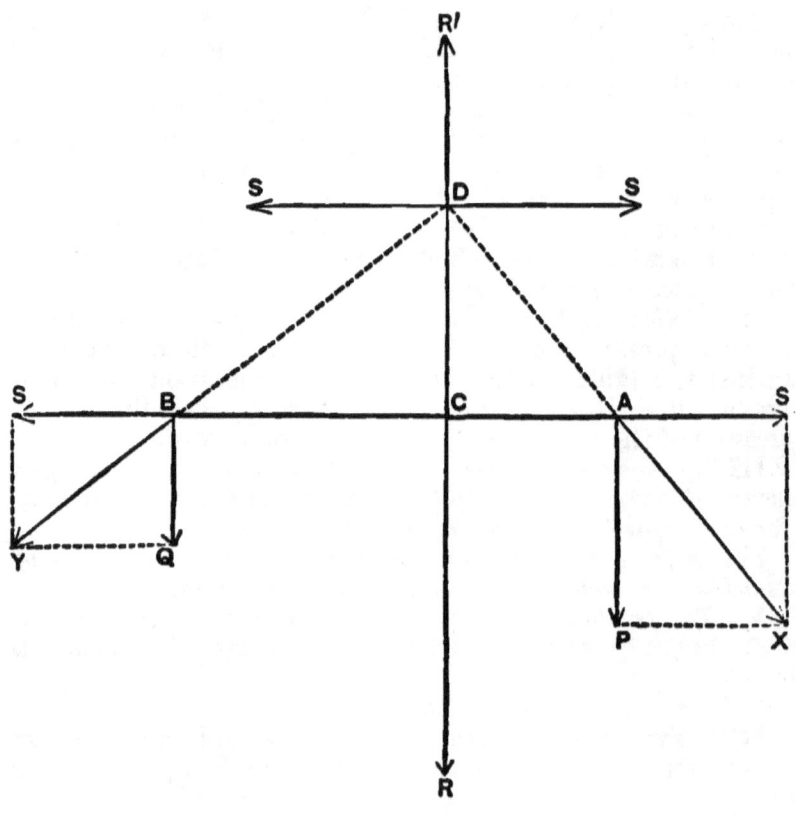

Fig. 6.

Hence the resultant of the forces P and Q is $P + Q$, and acts along a line D C parallel to the lines of action of P and Q in the same direction; so that it may be supposed to act at C. A force represented by C R' equal and opposite to C R is the balancing force.

ANSWERS AND SOLUTIONS.   145

To find the position of C. The triangles A P X and D C A are similar, being equiangular with respect to each other, and the line A S = the line P X.

Therefore $\dfrac{P}{S} = \dfrac{DC}{AC}$. Similarly $\dfrac{S}{Q} = \dfrac{BC}{DC}$. Therefore $\dfrac{P}{Q} = \dfrac{BC}{AC}$.

190. This theorem may be demonstrated in precisely the same way as the preceding, only the changed conditions require a separate figure. The student should draw the figure and go through with the demonstration.

Or, we may reason thus, using Fig. 6 and its conditions: The forces P, Q, and $R' = P + Q$, form a system of three parallel forces in equilibrium. Of these P and R' form a pair of unlike parallel forces held in equilibrium by Q, which is therefore by definition their balancing force (see beginning of solutions of exercises upon this Lesson); and a force equal to Q, and acting in the opposite direction at the point B, is their resultant. We have proved that $P : Q = BC : AC$. By theory of proportions, $P + Q : P = (BC + AC)$ or $AB : BC$, which proves the proposition.

193. For it is evident that the two preceding proofs hold equally good, whatever be the angle between the lines of action of the forces and the line drawn across them. In Fig. 6 this angle is a right angle, but this is not necessary.

194. Let P, Q, S, T, be the parallel forces acting at the points A, B, C, D, respectively. Join A B, and find a point L on A B such that $P : Q = BL : AL$. Then L is the point of application of the resultant of P and Q, which resultant is equal to $P + Q$. Join L C, and in the same way find M the point of application of $P + Q$ acting at L and S acting at C; this resultant will be $P + Q + S$. Proceeding in the same way with T we find N the point of application of $P + Q + S + T = R$, the resultant of the four

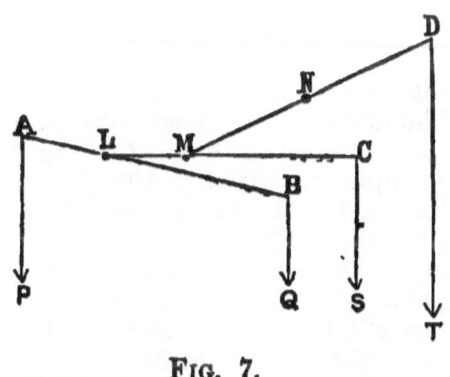

Fig. 7.

parallel forces. If any of the forces act in the opposite direction, the process of **190** must be used.

195 If the parallel forces in Fig. 7 were all turned through the same angular distance about their points of application in such a manner that their lines of action remained parallel, it can be easily seen that the point of application N of their resultant would remain unchanged, and the resultant would be turned about N through the same angular distance. This is true whatever be the number of the forces, and this remarkable constancy in position of the point of application of the resultant of any number of parallel forces has caused it to be called the *centre* of the system of parallel forces.

198. A *couple* is a system of two equal, unlike, parallel forces. Its arm is the distance between the lines of action of the two forces. Its *moment* is the product of either force into the arm, and is usually considered positive or negative according as it tends to produce rotation opposite to, or the same as, that of the hands of a watch. Thus in Fig. 8, P, P, are the forces, A B the arm, $P \times A B$ the moment which in this case is negative.

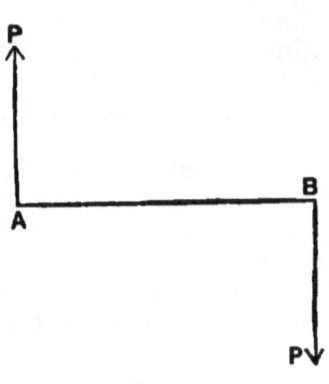

FIG. 8.

199. The *moment* of a force with respect to a point is equal to the product of the force into the distance from the point to the line of action of the force. The distinction between positive and negative moments is the same as that between positive and negative couples. For example (Fig. 9), P is the force acting in the direction O P. Then its moment with respect to the point A is $P \times A B$; its moment with respect to A$'$ is $P \times A' B'$; its moment with respect to A$''$ (or any point on its line of action) is $P \times 0 = 0$.

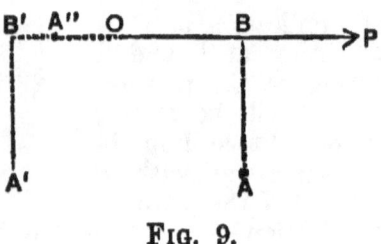

FIG. 9.

The doctrine of moments holds a position of the first importance in the mechanics of rigid bodies, for the reason that a moment is the appropriate measure of the tendency of a force to produce rotation.

200. "The moment of a force round any axis is the moment of its component in any plane perpendicular to the axis, round the point in which the plane is cut by the axis. Here we imagine the force resolved into two components, one parallel to the axis, which is ineffective so far as rotation round the axis is concerned; the other perpendicular to the axis, that is to say, having its line in any plane perpendicular to the axis. This latter component may be called the effective component of the force with respect to rotation round the axis. And its moment round the axis may be defined as its moment round the nearest point of the axis, which is equivalent to the preceding definition." (Thomson and Tait.)

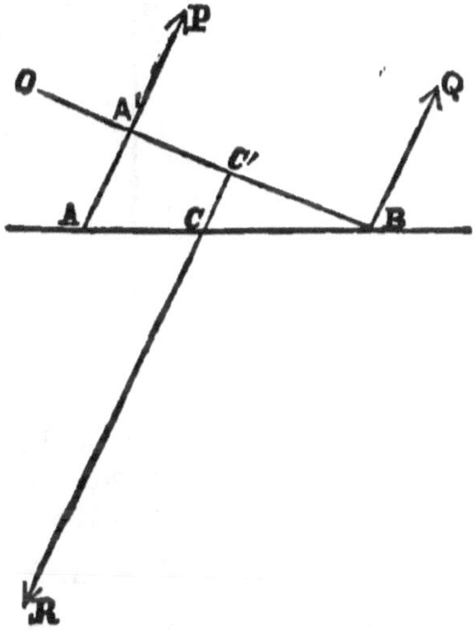

201. Let (Fig. 10) P, Q, and R $(= P + Q)$ be the forces acting at the points A, B, C, respectively. Choose any point O in the plane of the forces; join O B, the line O B cutting the lines of P and R at A' and C'. Then we have

FIG. 10.

$$R \times OC' = (P + Q) \times OC',$$
$$= P \times (OA' + A'C') + Q \times OC',$$
$$= P \times OA' + Q \times OB,$$

since, by **193**, $P \times A'C' = Q \times BC'$.

Therefore, $P \times OA' + Q \times OB - R \times OC' = 0$,

the first side of which equation is the algebraic sum of the moments of the forces.

202. Let P_1, P_2, P_3, &c., be the forces. Draw any line across their lines of action, and in it choose a point of reference, O. Let $O\,A_1 = a_1$, $O\,A_2 = a_2$, $O\,A_3 = a_3$, &c.

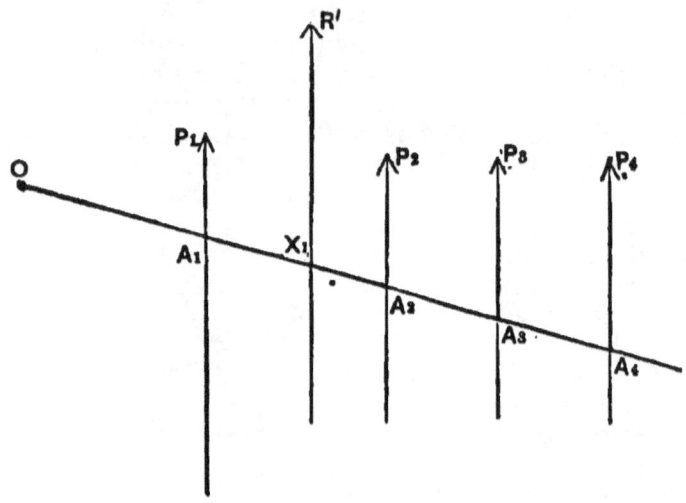

Fig. 11.

First find by means of **193** the resultant of P_1 and P_2; if we denote it by R', we have $R' = P_1 + P_2$. Divide $A_1\,A_2$ into parts inversely as the forces, so that

$$P_1 \times A_1\,X_1 =_2 P \times A_2\,X_1.$$

If we denote $O\,X_1$ by x' we have

$$P_1 \times (x' - a_1) = P_2 \times (a_2 - x'),$$

or, $(P_1 + P_2)\,x' = P_1\,a_1 + P_2\,a_2,$

that is, $R'\,x' = P_1\,a_1 + P_2\,a_2.$

Similarly we should find the resultant of R' and P_3 to be $R'' = P_1 + P_2 + P_3$, and that

$$R''\,x'' = R'\,x' + P_3\,a_3 = P_1\,a_1 + P_2\,a_2 + P_3\,a_3.$$

Hence, finally we have the two equations given in the statement of the Exercise.

Negative forces or negative values of any of the quantities a_1, a_s, &c., are included in this method, provided the generalized rules of multiplication and division in algebra are followed.

203. The necessary and sufficient conditions for the equilibrium of any number of parallel forces in one plane are the following:—

(1) algebraic sum of the forces = 0.
(2) algebraic sum of the moments of the forces round any point in the plane = 0.

If equation (1) does not hold, but equation (2) does, the forces have a single resultant whose line of action passes through the origin of co-ordinates.

If equation (1) does hold, but equation (2) does not hold, the system reduces to a couple; this is indicated by the fact that x is equal to infinity (since $R\,x$ is a finite quantity, and R by hypothesis is zero); in other words, the point of application of the resultant is at an infinite distance, which is the case with a couple.

204. In the first kind of Levers, the fulcrum is between P and W; in the second kind, W is between P and the fulcrum; in the third kind, P is between W and the fulcrum.

207. The *mechanical advantage* of any mechanical contrivance or combination whatsoever is expressed by the fraction $\frac{W}{P}$, P being the applied force or power, W the weight which is raised or pressure which is exerted, and friction, &c., being neglected. In the case of the lever this ratio is often called by workmen the *leverage*. There exists a popular impression, that a machine can generate or create force. This impression arises from the well-known fact that by the aid of a machine an enormous weight can be raised, or resistance overcome, by the application of a very small power. But on examination it will be found that in the exact proportion in which a machine diminishes the power required to raise a given weight, it increases the distance through which this power must act in order to raise the weight through a given height. Take a lever the arms of which are as 10 to 1; we have seen that a pound weight hung at the extremity of the longer arm will balance 10 pounds at the end of the shorter arm; the slightest addition to the pound weight, or rather, friction being neglected, the slightest exterior impulse, as a touch with the finger, will cause the pound weight to descend and raise the 10 pounds. But a little geometrical reflection will make it obvious that, in order that the 10 pounds may rise 1 inch, the pound weight must descend 10 inches. A more complex yet familiar instance is furnished by the mechanism of a watch. Here the power communicated by the main-spring is applied to a train of wheels, and produces a much more rapid movement in the balance-wheel. But the slightest touch of the finger will check the balance-wheel, while the winding up of the main-spring requires a far greater force. And on examination it would be found that, making allowance for friction, the force with which the balance-wheel moved was precisely as many times less than that required to move the main-spring as its motion was more rapid. In fact, one grand and simple law is rigorously fulfilled by every machine and mechanical combination, however complex. This law, if we neglect friction, may be stated as follows: —

In every machine the product of the power and the distance through which the power moves in its own direction is equal to the product of the weight (*or resistance overcome*) *and the distance through which the weight moves in its own direction.*

This is a modified form of the celebrated principle of Virtual Velocities; it may now be more appropriately termed the Principle of Work. In fact, adopting the idea and measure of work, explained by Mr. Stewart in Chapter III., this principle is stated simply by saying that, *friction being neglected, the work transmitted by a machine is unaltered in amount;* or more generally, friction being included, *in every machine, the parts of which are moving uniformly, the work done by the power = the work done against the resistance + the work done against friction.*

Thus it appears that a machine may increase or diminish the *magnitude* of the power in any given ratio, but that it cannot increase the product of the force acting at any part into the space through which this force moves, or, in other words, it cannot increase the work transmitted by the machine. In point of fact, this product is diminished by the effect of friction as the force is transmitted through the machine. The use of a machine is to apply more advantageously the force applied to it, to transmit force, and to change in a desired degree the direction and velocity of motion.

211. Let a and b denote the unequal arms, and x the true weight of the body. Put the body on the pan attached to the arm a, and balance it by weights in the other pan; call these weights W. Next reverse the operation by putting the body on the pan attached to the arm b, and balancing it by weights W' on the pan attached to the arm a. Then W and W' are the two false weights. Now in the first operation we have, by **205**,—
$$a x = W b,$$
and in the second operation similarly we have
$$b x = W' a.$$
Hence $\qquad a b x^2 = W W' a b,$
or $\qquad x^2 = W W'.$
And $\qquad x = \sqrt{W W'}.$

215. The Wheel and Axle is essentially a Lever of the first kind, the fulcrum being the centre of the axle, and the power and weight being applied at the extremities of radii of the

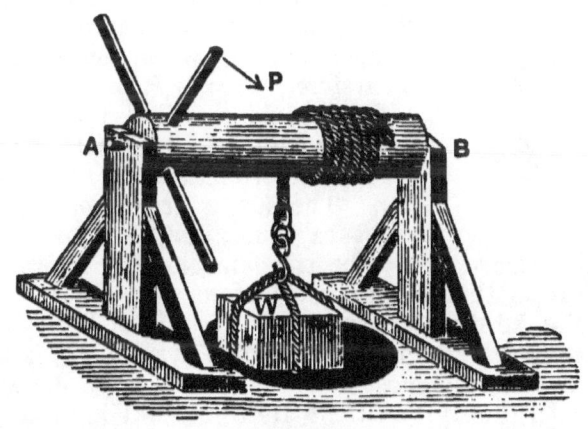

FIG. 12.

wheel and the axle respectively. This is easily seen by an inspection of Fig. 12, which represents a common Windlass.

Here in place of a wheel we have what is equivalent, four arms or spokes, which are turned by hand. The proportion given in the statement of the Exercise can be readily obtained by the student.

The Wheel and Axle possesses an important advantage over the simple Lever, in the fact that it allows us to raise a weight through any given height. With a simple Lever a weight can be raised only a small distance before it becomes necessary to place the Lever in a new position, and to support the weight by some other force while this change is being made. The Wheel and Axle is a practical arrangement for continuing the action of a Lever as long as may be required, the weight rising all the time.

217. One simple principle will explain the mechanical advantage of any system of Pulleys, however complex, namely, the principle of the tension of a cord, given in the solution of **169**. The single movable Pulley is shown in Fig. 13. There

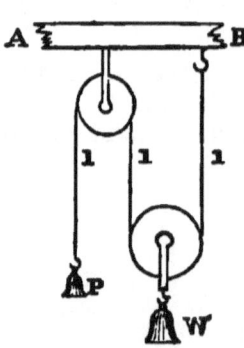

FIG. 13.

is only one cord, which, by the principle referred to, has the same tension throughout (friction, &c., being neglected). The weight, W, is equally supported by the parts of this cord in contact with the Pulley at A and B. Therefore $W = 2 \times$ tension of the cord. But it is evident that $P =$ tension of the cord. Therefore, $W = 2P$.

The single movable Pulley may also be regarded as a Lever with equal arms, and the fulcrum at the centre of the Pulley. The forces acting on equal arms must be equal, and therefore the pressure on the fulcrum is twice the tension of the cord, or $2P$. But W causes this pressure; therefore, $W = 2P$.

218. Fig. 14 represents the three systems of Pulleys referred to in Exercises **218–220**. Of these systems (1) is the simplest, and also the most convenient in use. We will solve case (2). In this system there are in general n (here 5) movable Pulleys, each hanging by a separate cord as shown in the figure. The tension of the cord which passes under the highest movable Pulley A is by the principle of tension equal to the power P. By the same reasoning which was used in the case

ANSWERS AND SOLUTIONS. 153

of the single movable Pulley, the tension of the cord passing under the next lower Pulley is $2P$, of the cord passing under the third Pulley is $4P = 2^2 P$, of the next cord is $2^3 P$, of the cord passing under the lowest Pulley is $2^4 P$. The tension of this last cord is by the same reasoning equal to $\frac{1}{2} W$.

Therefore, $W = 2^5 P$. And, in general, $W = 2^n P$.

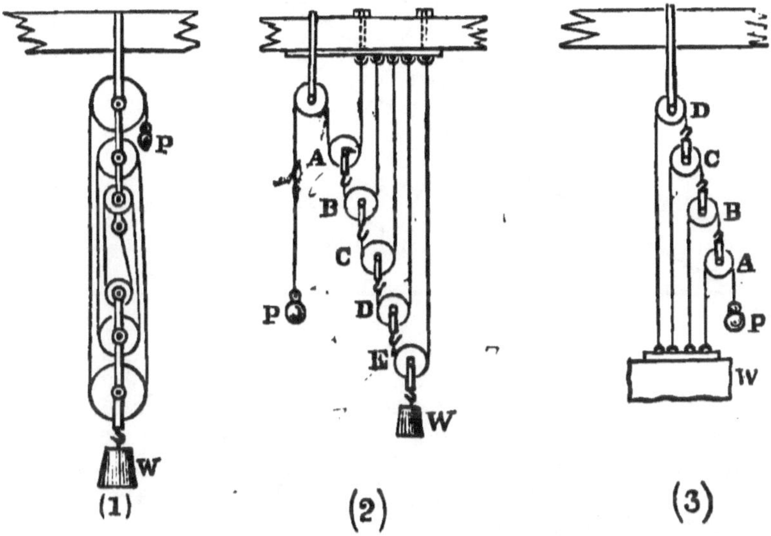

Fig. 14.

223. The Inclined Plane is a plane inclined at any angle to the horizon. The principle of the Inclined Plane consists in this, that a weight W can be supported on the Inclined Plane by a power P which is less than W. It is a direct example of the Parallelogram of Forces.

Conceive that the Plane is perfectly smooth, and likewise the body which is resting on the Plane. This body has its centre of gravity at G, so that, as will be explained later, its weight W may be regarded as entirely concentrated in a heavy point at G. Three forces are in action when the body is in equilibrium, the weight of the body W acting at G vertically downwards, the power P applied to keep the body at rest on the Plane and acting from G up the Plane, and the reaction R of the Plane arising from the pressure of the body

154   ELEMENTARY PHYSICS.

on the Plane and acting in a direction perpendicular to the Plane. Substitute for W its two components G M and G N,

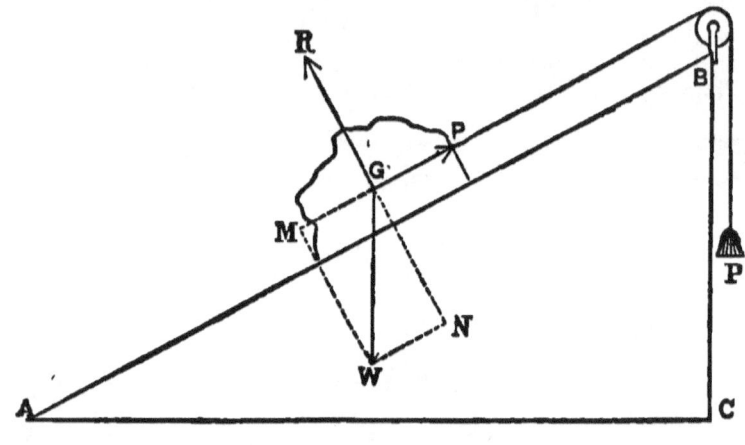

FIG. 15.

parallel and perpendicular respectively to the Plane. These components are opposite to P and R respectively; hence they must be equal to them each to each, or, G M $= P$, and G N $= R$. The triangles A B C and G W N are similar; whence it follows that

$$\frac{P}{W} = \frac{BC}{AB} = \frac{\text{height}}{\text{length}},$$

and that

$$\frac{R}{W} = \frac{AC}{AB} = \frac{\text{base}}{\text{length}}.$$

226. Let the student represent the conditions of this question by a diagram similar to that given in the solution of **223**. He will see that the lines representing the three forces P, W, and R form a right triangle. This triangle may be shown to be similar to the large triangle the hypothenuse of which is the Plane. Hence the required proportion is easily obtained.

227. The Screw is a movable inclined plane wrapped round a right cylinder. The relation between the Inclined Plane and the Screw may be understood by reference to Fig. 16.

ANSWERS AND SOLUTIONS. 155

Suppose that we unroll paper previously rolled round the right cylinder, A B K L, by causing the cylinder to revolve on its axis through exactly one revolution; we shall obtain a

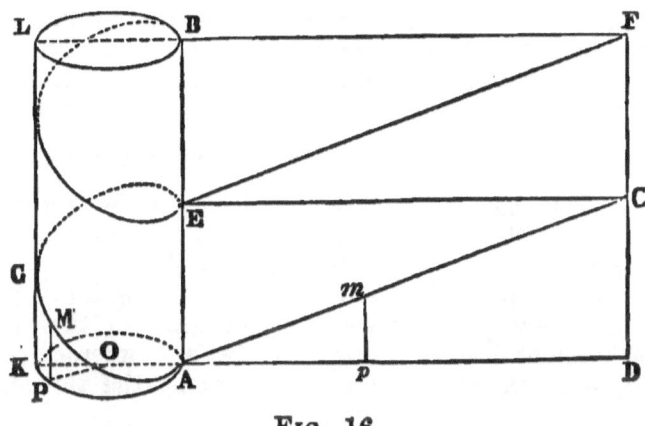

FIG. 16.

rectangle, B A D F. Divide this rectangle into a certain number of equal rectangles by drawing lines parallel to the base, A D; draw diagonals to these rectangles as shown in the figure; and lastly, roll the entire rectangle again upon the cylinder. The diagonals will form a continuous curve called a *helix*, composed of spirals, one above another, equal in number to the number of partial rectangles (in this case two).

In rolling up the rectangle the line mp for example falls into the position M P, the point D falls, after one revolution, upon A, the point C falls upon E, and F upon B. If, now, we take a cylinder (of wood or iron) having on its surface, in place of a helix, traced in pencil or ink, a projecting *thread* making at all points the same given angle (called the *pitch* of the Screw) with the horizon, we have a Screw ready for use.

Thus, the Inclined Plane A C forms one revolution of the helix, its base is equal to the circumference of the cylinder, and its height to the *distance between two threads*, a quantity of special importance in the theory of the Screw.

228. In order to adapt the Screw to use, a *nut* must be employed; this is a block pierced with an equal cylindrical aperture, upon the inner surface of which is cut a groove the exact counterpart of the thread of the Screw. It is evident that the Screw can only be made to move in the nut by revolving about its axis. Suppose the Screw, with its axis vertical, to be contained in the nut without friction, and let a weight, W, be placed upon it; then the pressure due to this weight will be transmitted to every point of the thread of the Screw. Suppose that there are n points on the thread, each point in contact with a corresponding point of the groove in the nut; then each point is acted upon by a vertical force equal to $\dfrac{W}{n}$.

Now it is clear that under these circumstances the Screw would descend by revolving on its axis (that is, the various points of the thread would slide down the groove with a spiral motion), unless prevented by some force; call this force P_1, and let it act at some point on the circumference of the cylinder along a horizontal line tangent to the circumference. This force P_1 will likewise be distributed along the thread, each point being acted upon by a horizontal force equal to $\dfrac{P_1}{n}$. Consider now the equilibrium of any point A of the thread. It is kept at rest by three forces, $\dfrac{P_1}{n}$, $\dfrac{W}{n}$, and the reaction of the thread, which is equal and opposite to the normal pressure on the thread arising from the combined action of $\dfrac{P_1}{n}$ and $\dfrac{W}{n}$, and which need not be further considered in this investigation. Bearing in mind the analogy between the Screw and the Inclined Plane, it is evident that this case is similar to that treated in **226**, and considered as an instance of the Inclined Plane, we have, $\dfrac{P_1}{n} : \dfrac{W}{n}$ = height of plane : base of plane. Simplifying the first ratio, and substituting for the terms of the second ratio their equivalents in terms of the Screw, we obtain $P_1 : W$ = distance between two threads : circumference of the cylinder.

Practically, the Screw is never used as a simple machine, the power being applied by means of a Lever passing either through the head of the Screw or through the nut. The Screw acts, therefore, with the combined power of the Lever and the Inclined Plane; and in investigating the effect we must take

ANSWERS AND SOLUTIONS.

into account both these Mechanical Powers. The proportion just given represents the effect of the Inclined Plane. To combine this with the effect of the Lever is left as an exercise for the student. The result is stated on page 91.

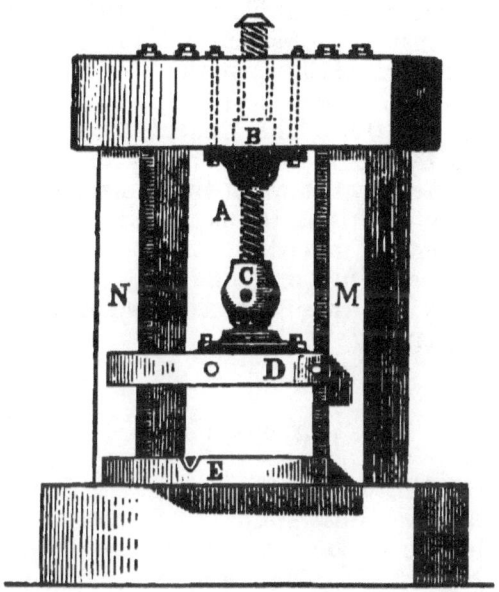

Fig. 17.

To produce pressure with the Screw either the Screw or the nut must be fixed; whichever is free is then turned, and made to press against the resistance. It is evident that one revolution causes the Screw (or the nut) to advance by an amount equal to the distance between two threads. In the common Screw-press, represented in Fig. 17, the nut is fixed and the Screw movable. A is the Screw, enlarged below at C, where it is pierced with two holes at right angles for applying the Lever. B is the nut firmly fixed in the upper part of the press. By turning the Screw in the nut it descends and pushes before it the plate D, which works in guides so that it can have only a vertical motion. The body is compressed by placing it on the fixed platform E, and lowering the Screw and plate D.

In deducing the law of equilibrium of the Screw we have neglected the effect of friction for the sake of simplicity. But the amount of friction in every Screw is very great; in fact, the Screw owes its utility to friction; for if there were no friction the Screw would *overhaul*, that is, turn backwards the instant the power was removed. To prevent this the friction must be sufficient to consume more than half the power applied.

Note. — All the Exercises upon the Mechanical Powers can be very readily solved by means of the principle of Work stated in the solution of **207**. The student should first work out these Exercises by the methods already given or indicated, and then solve them by the aid of the principle of Work. The greater simplicity of this latter method is strikingly shown in the case of the Screw; indeed, it is obvious that, friction being neglected, this beautiful principle may be employed with equal ease in any machine, whatever its nature and however complex its mechanism.

Lesson VI.

271. For the Third Law of Motion, see Solution of **135**. Newton gives three illustrations.

1. "If you press a stone with your finger, the finger is also pressed by the stone." This is an illustration of forces in equilibrium (see Solution of **169**).

2. "If a horse draws a stone tied to a rope, the horse (if I may so say) will be equally drawn back towards the stone; for the distended rope, by the same endeavor to relax or unbend itself, will draw the horse as much towards the stone as it does the stone towards the horse, and will obstruct the progress of the one as much as it advances that of the other."

3. "If a body impinge upon another, and by its force change the motion of the other, that body also (because of the equality of the mutual pressure) will undergo an equal change, in its own motion, towards the contrary part. The changes made by these actions are equal, not in the velocities, but in the quantities of motion (momenta) of the bodies."

272. "Every force acts between two bodies, or parts of bodies. If we are considering a particular body or system of bodies, then those forces which act between bodies belonging to this system and bodies not belonging to the system are called External Forces, and those which act between the different parts of the system itself are called Internal Forces." (*Maxwell*.)

The same force may play the part of an interior or exterior force according to the point of view adopted. Take, for instance, the motion of a body which falls to the earth; the attraction of one of the particles of this body for a particle of the earth is an exterior force; if, on the other hand, we consider the motion of the material system formed of the body and the earth, this same attraction becomes an interior force. The student should find other illustrations.

279. For this distinction, see the solution of **56**. The student should give illustrations. The duration of the action of an impulsive force is too brief to be measured, so that we cannot measure the force by the momentum generated in one second. An impulsive force is measured by *the total momentum which it generates*.

280. A body is said to be *perfectly elastic* if, when it impinges perpendicularly on a fixed plane, it will recoil back from the plane with equal velocity, or when the *velocity of recoil is equal to the velocity of approach*. An *imperfectly elastic* body is one whose *velocity of recoil is less than its velocity of approach*. A *perfectly inelastic body* is one whose *velocity of recoil is zero*. The ratio of the velocity of recoil to that of approach is called the *coefficient of restitution*.

281. The principle of conservation of momentum is stated in the enunciation of **292**. In applying it to the case supposed, we have for the material system the two equal bodies; and since their velocities are equal and in opposite directions, the algebraic sum of their momenta is zero. The collision does not alter this sum, for it brings both bodies to rest.

LESSON VII.

306. Let m, m' denote the masses, d their distance apart; then, —

$$\text{force} = \frac{m\,m'}{d^2}.$$

The simplest unit of attractive force is defined by putting $m = m' = d = 1$.

160 ELEMENTARY PHYSICS.

Lesson IX.

335. Regarding bodies as composed of particles, each acted on by gravity, Mr. Stewart has given in § 47 a definition of the centre of gravity of a body which may be stated in other words thus : — A point in the body such that, if supported, the body will remain at rest, but, if not supported, the body will fall under the action of gravity. But this definition is an obvious consequence of the following mechanical definition, which should be read in connection with the solution of **195**: *the centre of gravity of a body is the centre of the system of parallel forces exerted by gravity upon the body.*

336. A body is said to be *homogeneous*, from a mechanical point of view, when equal volumes of the body have equal masses. A *plane of symmetry* in a body is a plane such that every perpendicular line passing through it cuts the surface of the body in two points equally distant from the plane. A body may have more than one plane of symmetry; in a sphere, for example, every great circle is a plane of symmetry.

Similarly, if a line can be drawn through a body, such that every perpendicular drawn through the line cuts the surface of the body in two points equally distant from the line, this line is called a *line* (or *axis*) of symmetry, with respect to the body.

The surface of the body is said to be *symmetrical* with respect to the plane, or the line, as the case may be.

A point is called the *centre* of a surface when every straight line drawn through the point and terminated by the surface is bisected by the point. The centre of a surface is also the centre of the body which is bounded by the surface. Considered in this light, it is also called the *centre of figure*, or *geometric centre*.

340. In this, and the three next Exercises, the student should find and state the particular conditions of equilibrium for each case, and should define and distinguish between *stable*, *unstable*, and *neutral* equilibrium, and give examples of each kind.

The general principle to be applied in all cases is that the centre of gravity must be supported in order to have equilibrium.

348. The principal use of the pendulum in Physics is to determine the value of the acceleration due to gravity.

ANSWERS AND SOLUTIONS. 161

Taking the formula for the simple pendulum given in **360**, we easily find that $g = \dfrac{\pi^2 l}{T^2}$, whence it follows that the value of g can be found by making a pendulum vibrate, and measuring l and T. But to measure accurately these quantities is by no means a simple matter.

In the first place, the formula applies directly only to the *simple* pendulum, which is defined as a heavy particle suspended from a fixed point by an inextensible weightless thread. Such a pendulum can exist only in the mind, and is a conception employed by the investigator for the sake of simplifying the inquiry after the laws of pendulum oscillations. In practice we employ bodies which oscillate about a horizontal axis, called the *axis of suspension;* these are termed, for the sake of distinction, *compound pendulums*. The length of a compound pendulum is the length of a *synchronous* simple pendulum, that is, of a simple pendulum which will oscillate in the same time as the given compound pendulum. This length can be calculated in a given case, when the pendulum is of regular form, by the aid of formulae which are given in higher treatises on Dynamics; but its value is more easily obtained by what is called Kater's method.

This method is founded on a remarkable property of a compound pendulum, discovered by Huyghens, called the *convertibility of the axes of suspension and oscillation*. The axis of oscillation is an axis parallel to the axis of suspension in the plane containing it and the centre of gravity of the pendulum, and at a distance from it equal to the length of the synchronous simple pendulum. The body, in fact, oscillates as if its entire mass were collected on the axis of oscillation. Now the property discovered by Huyghens was this: that if we suspend a pendulum by its axis of oscillation, the former axis of suspension becomes the new axis of oscillation, and the pendulum oscillates in the same time as before. Kater constructed a reversible pendulum, which could be supported by either of two parallel knife-edges, one of which could be adjusted to any distance from the other. The length of this pendulum could be found by the method of repeated trials with a great degree of accuracy.

In order to measure T, the method which naturally suggests itself is to count the number of oscillations which take

K

place in a given time, and then divide the time by the number of oscillations.

This is, however, far from an easy process, if accuracy in the results is aimed at. Borda devised a great improvement upon this method, by comparing the motion of the pendulum with the motion of the pendulum of an astronomical clock regulated to beat seconds. By the use of Borda's method, called the "method of coincidences," one can calculate the number of oscillations without being obliged to count them.

Another method of finding, by the aid of a pendulum, the value of g consists in employing a seconds pendulum of regular form; in this case $T = 1$, and l can be calculated in the manner already stated.

368. Let P denote the force or pressure perpendicular to two surfaces in contact, by which the surfaces are pressed together; and let F denote the *least* force parallel to the surfaces in contact which is able to move one surface along the other: then, the ratio of F to P is called the *coefficient of friction* for the two surfaces.

If a body be placed on an inclined plane whose angle of inclination can be altered at pleasure, then that inclination of the plane for which the body is just about to slide is called *the angle of friction* for the materials of which the body and the plane are composed.

370. There are five forces to be considered: (1) Gravity, which may be regarded as acting entirely at the centre of gravity of the ladder, (2) and (3) the reactions of the floor and wall perpendicular to the surface in each case, (4) and (5) the forces due to friction, and which act parallel to the surface of the floor and wall.

371. Referring back to Fig. 15, and using the same notation, it is plain that when the body is on the point of sliding, the friction F must be equal to the component of W, which is parallel to the plane, that is to $W \dfrac{\text{height}}{\text{length}}$. The pressure R on the plane is equal to $W \dfrac{\text{base}}{\text{length}}$. Therefore we have

$$\text{the coefficient of friction} = \frac{F}{R} = \frac{\text{height}}{\text{base}}.$$

Lesson XI.

376. A *perfect fluid* is an ideal conception, like that of a rigid or smooth body; it is defined as *a body incapable of resisting a change of shape*. Common liquids and gases fulfil this definition when in a state of rest, but no existing fluid fulfils the definition when it is in motion.

"All actual fluids are imperfect, and exhibit the phenomenon of internal friction or *viscosity*, by which their motion after being stirred about in a vessel is gradually stopped, and the energy of the motion converted into heat." (*Maxwell.*)

A *viscous* fluid is a substance such that the very smallest force applied to it will produce a constantly increasing change of form. The change of form may take place very slowly; but if it takes place so as to be sensible, and continually increases with the time, the substance is viscous.

"Thus a block of pitch may be so hard that you cannot make a dint in it by striking it with your knuckles; and yet it will, in the course of time, flatten itself out by its own weight, and glide down hill like a stream of water." (*Maxwell.*)

378. It follows from the foregoing definition of a perfect fluid, that its pressure on any surface must be at all points perpendicular to the surface. The student should be able to give the reason.

379. Let S, S', denote the two surfaces, and P, P', the pressures; then $S : S' = P : P'$. Pascal's principle cannot be proved directly, on account of the action of the force of gravity.

385. The proof required is simply this: if the resultant of all the forces which act on the liquid at any point of its surface were not normal (*i. e.* perpendicular) to the surface at that point, then resolve the resultant into components perpendicular and parallel to the surface. The first component is neutralized by the resist-

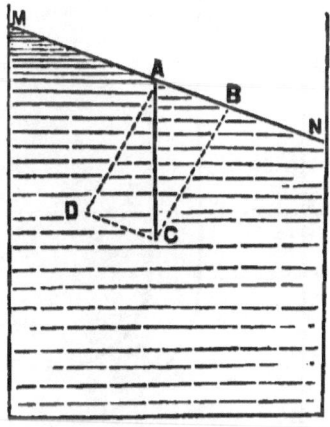

Fig. 18.

ance to compression of the liquid, liquids being practically incompressible; the other component would cause the particle on which it acts to move along the surface, which is contrary to the supposition that the liquid is at rest.

Thus, in Fig. 18, let A C represent this resultant. Then A D and A B are the components referred to, and it is easy to see that the component A B being unresisted would cause motion.

386. The surface is everywhere termed *horizontal*. Small extents of the earth's surface may be considered planes, but large areas must be regarded as nearly spherical.

387. (1) follows from Pascal's principle, (2) is self-evident, and (3) is likewise self-evident when the meaning of density is taken into account.

388. By **387** the pressure is proportional to area $\times$ depth $\times$ density. Now if the centimetre be used as a unit of length, this product is by **22** the weight in grammes of a volume of the liquid, having the given area and depth.

Adopting the unit of weight as the unit of hydrostatic pressure, this product is the pressure in grammes.

399. Let d, d', denote the densities of the two liquids, d' being greater than d. Let C D and A B (Fig. 19) be horizontal lines drawn through the sur-

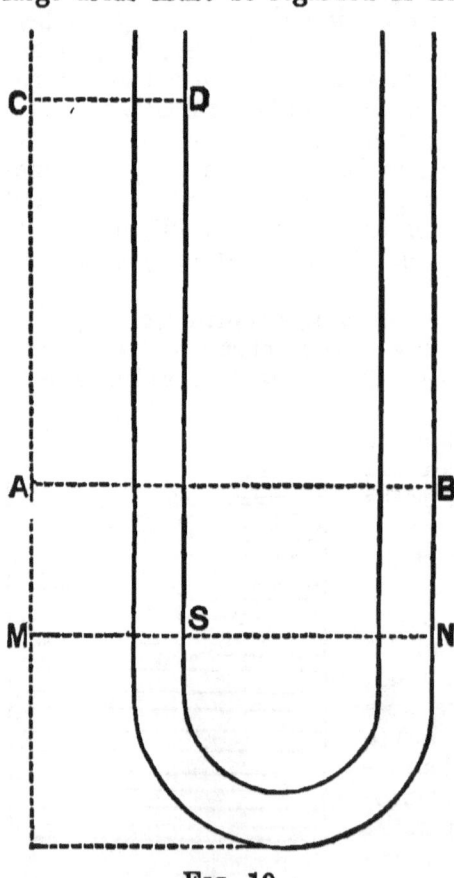

FIG. 19.

faces of the two liquids at D and B, the surface of the lighter liquid being at D. Let M N be a horizontal line drawn through the common surface of contact of the liquids at S, and let C A M be a vertical line. Finally, let s, s', denote the areas of sections of the tube at S and N respectively, and put $CM = h$, and $AM = h'$. Since the liquid columns are in equilibrium, it follows from **387** (1) that the pressures exerted upon the surfaces s and s' by the superincumbent liquids must be as the areas of the surfaces. The pressure upon $s = s\,h\,d$; that upon $s' = s'\,h'\,d'$. Therefore

or
$$s\,h\,d : s'\,h'\,d' = s : s'$$
$$h\,d = h'\,d'$$
that is, $$h : h' = d' : d$$

400. The pressures against the various points on the vertical side of the vessel form a system of parallel forces, each force being proportional to the depth of the point to which it is applied. This condition of things is partially represented by the arrows drawn in Fig. 20.

Fig. 20.

166 ELEMENTARY PHYSICS.

A very little reflection is sufficient to show that the point of application of the resultant of this system must be somewhere *below* G, the centre of gravity of the side. This becomes evident by considering that all the pressures below G are greater than any of the pressures above G. Calculation shows that this point of application for a rectangular surface is at a point C, just two thirds the depth of the side. This point is called the *centre of pressure* of the side.

405. For the definition and measure of *density*, see Solution of **21**. The *specific gravity* of a substance is the *ratio of its density to that of some standard substance*, usually water at its maximum density.

The density of water in the Metric System being unity, it follows that in this system the density and specific gravity of a substance are numerically the same.

422. The method commonly employed is to attach to the body a *sinker*, that is, a body heavier than water, and large enough to cause both bodies to sink. The specific gravity of the sinker being supposed known, it is easy to determine that of the body; the student should deduce a formula for this purpose.

Lesson XII.

452. The three forms referred to are as follows :—

1. By the number of statical units of force in the pressure on unit of area ; *e. g.* in pounds' weight for square inch, or in kilogrammes' weight per square centimetre. Pressures thus expressed are reduced to dynamical measure by multiplying by the value of g at the given locality.

2. By the height of a column of mercury at 0° C. which would exert by its weight an equal pressure ; thus the average pressure of the atmosphere is described as a pressure of 30 inches, or 76 centimetres, of mercury.

3. In terms of a large unit which is nearly equal to the average atmospheric pressure at the level of the sea. This unit is called an *atmosphere*, and is used chiefly in measuring pressures in boilers, and in scientific experiments which require very great pressures.

These three measures are thus related : in the British system, one atmosphere = pressure due to a height of 29·905

ANSWERS AND SOLUTIONS. 167

inches of mercury at 32° F. at London, where the force of gravity is 32·1889 feet = about 14¾ lbs. weight per square inch. In the Metric system one atmosphere = pressure due to a height of 76 centimetres of mercury at 0° C. at Paris, where the force of gravity is 9·80868 metres = about 1033 grammes' weight per square centimetre.

One British atmosphere = 0·99968 of a Metric atmosphere.

455. The volume of the mercury resting upon unit section of the tube at the bottom of the barometer is h, and its mass, and also its weight, is $13·596\,h$. This is reduced to dynamical measure by multiplying by g.

460. In such questions as this the effect of any increase in the atmospheric pressure on the density of water is too small to be considered.

464. Let σ = specific gravity of the air at the sea-level, ρ = that of mercury, h = height of barometer at the sea-level. Then the pressure per unit area indicated by the barometer is $h\rho$. If x = height of the homogeneous atmosphere, the pressure produced by it per unit area is $x\sigma$. Hence, $x\sigma = h\rho$; or, $x = \dfrac{\rho}{\sigma} h$. To find x in metres, substitute for h 76 centimetres, for ρ 13·596, and for σ $\tfrac{1}{773}$.

473. No account is to be taken of the effect of increased pressure on the density of the water.

478. Let V and v denote the volumes of the receiver and the barrel respectively, D_0 the initial density of the air, D_1, D_2, D_3, &c., ... D_n, the densities after 1, 2, 3, &c., ... n, strokes of the piston. When the piston is first raised the volume V of air becomes increased to $V + v$. Hence by Boyle's law, —

$$D_0 : D_1 = V + v : V,$$

or,
$$D_1 = D_0 \frac{V}{V + v}.$$

Similar reasoning shows that

$$D_2 = D_1 \frac{V}{V + v} = D_0 \left(\frac{V}{V + v}\right)^2,$$

and thus finally we have

$$D_n = D_0 \left(\frac{V}{V + v}\right)^n.$$

479. One limit is pointed out by Mr. Stewart at the close of § 93. Since each stroke of the piston only removes *a part* of the air which remains, it is plain that the receiver can never be completely exhausted. This is indicated by the formula of the last Exercise; as n increases, the value of D_n diminishes towards zero as a limit, but this limit is reached only when n is made greater than any assignable quantity.

There are other causes which tend to produce an actual limit after a definite and not very large number of strokes.

In the first place, it is impossible to avoid the existence of a small space between the bottom of the barrel and the bottom of the piston, when the latter is in its lowest position. This space has been called *untraversed space;* and it is evident that it contains a small quantity of air at the atmospheric pressure, when the piston is in its lowest position. When the piston is drawn up, this air is rarefied; but unless its tension becomes less than the tension of the air remaining in the receiver, no air can flow from the receiver into the barrel, and the pump ceases to produce any effect.

A second cause is *leakage*, which exists in the best-constructed air-pumps, and which increases rapidly as the tension of the air in the receiver and barrel diminishes. As the piston descends it expels a certain quantity of air; as it ascends a certain quantity of air enters from leakage. If these two quantities are equal, the limit of rarefaction has evidently been reached.

Lastly, perhaps the most serious cause is the *absorption of air by the oil* used in lubricating the piston. This oil is poured upon the top of the piston, where it is forced by the pressure of the external air between the piston and the sides of the barrel, and finally falls to the bottom of the barrel. Here it absorbs air which it gives out in part during the ascent of the piston. Hence arises another limit to the degree of rarefaction.

APPENDIX.

(*Containing useful Data, Tables, and Formulæ.*)

I.

ENGLISH WEIGHTS AND MEASURES.

THE fundamental units of Time, Length, and Mass, respectively, are the Mean Solar Day, the Imperial Standard Yard, and the Imperial Standard Pound Avoirdupois.

The Mean Solar Day is the *average* interval between two successive passages of the sun across the meridian. The mean solar day is divided into 24 *hours*, each hour into 60 *minutes*, and each minute into 60 *seconds*, so that one second is $\frac{1}{86400}$ part of a day. For a great number of purposes the mean solar day is an inconveniently large unit, and the minute or second is therefore employed.

The Imperial Standard Yard is, by Act of Parliament, the distance between two points in a certain bronze bar deposited in the Office of the Exchequer in London, the temperature of the bar being 62° F. (see 18 and 19 Vict. c. 72, July 30, 1855).

The Imperial Standard Pound Avoirdupois is, by the Act above cited, a platinum weight marked "P. S. 1844, 1 lb.," deposited in the Office of the Exchequer in London.

TABLE I.

Measures of Length.

	in.	ft.	yd.	rd.	fur.	m.
Inch	1					
Foot	12	1				
Yard	36	3	1			
Rod	198	16½	5½	1		
Furlong	7 960	660	220	40	1	
Mile	63 360	5 280	1 760	320	8	1

APPENDIX.

NOTES. 1. In scientific investigations the foot and the inch are generally employed, being more convenient than the yard. The inch is subdivided decimally and also binarily (*i. e.* into halves, quarters, eighths, etc.).

2. The mile in the above table is the *statute* mile. The geographical or nautical mile (also called knot) is $\frac{1}{60}$ of a degree of longitude on the equator or $\frac{1}{21600}$ of the whole equator, and is rather more than 1·15 statute miles.

3. A *hand* = 4 inches; 1 *fathom* = 6 feet; 1 *league* = 3 miles.

TABLE II.

Measures of Surface.

	sq. in.	sq. ft.	sq. yd.	sq. rd.	R.	A.	sq. m.
Square Inch	1						
Square Foot	144	1					
Square Yard	1 296	9	1				
Square Rod	39 204	272¼	30¼	1			
Rood	1 568 160	10 890	1 210	40	1		
Acre	6 272 640	43 560	4 840	160	4	1	
Square Mile	4 014 489 600	27 878 400	3 097 600	102 400	2 560	640	1

TABLE III.

Measures of Volume.

	cub. in.	cub. ft.	cub. yd.
Cubic Inch	1		
Cubic Foot	1 728	1	
Cubic Yard	46 656	27	1

NOTE. 16 cubic feet of wood make 1 cord foot, and 8 cord feet or 128 cubic feet make 1 *cord*.

APPENDIX.

TABLE IV.

Measures of Mass (Avoirdupois Weights).

	gr.	dr.	oz.	lb.	qr.	cwt.	T.
Grain	1						
Drachm	27·34375	1					
Ounce	437·5	16	1				
Pound	7 000	256	16	1			
Quarter	196 000	7 168	448	28	1		
Cwt.	784 000	28 672	1 712	112	4	1	
Ton	15 680 000	573 440	34 240	2 240	80	20	1

NOTES. 1. The pound is connected with the unit of volume as follows: 1 cubic inch of distilled water weighed in air at 62° F. (bar. 30 inches) = 252·458 grains.

2. The Avoirdupois Pound of matter is equal to the mass of 27·7274 cubic inches of distilled water, weighed as above.

3. For many purposes it is sufficiently accurate to call the weight of 1 cubic foot of water 1000 ounces.

Three other measures are much in use, viz., Liquid Measure, the base of which is the Imperial Gallon; Dry Measure, the base of which is the Imperial Bushel; and Troy Weight, the base of which is the Troy Pound.

The Imperial Gallon is a measure of volume or capacity which will contain 10 pounds, avoirdupois weight, of distilled water, weighed in air, at 62° F., the barometer at 30 inches.

The Imperial Bushel is also a measure of capacity, and is equal to 8 imperial gallons.

The Troy Pound is a measure of mass, bearing to the Avoirdupois Pound the ratio of 5760 : 7000; *i.e.* it contains 5760 grains' weight of matter.

APPENDIX.

TABLE V.
Measures of Capacity.

Liquid Measure.	Dry Measure.
4 gills = 1 pint (pt.) 2 pints = 1 quart (qt.) 4 quarts = 1 gallon (gal.) 52½ gallons = 1 hogshead.	2 pints = 1 quart. 8 quarts = 1 peck (pk.) 4 pecks = 1 bushel (bush.)

NOTES. 1. Liquid Measure is used in measuring liquids, and Dry Measure in measuring solid matter consisting of small parts or pieces, as grain, fruit, salt, roots, ashes, &c.

2. The *tun* = 2 *pipes* = 4 hogsheads = 210 Imperial gallons.

TABLE VI.
Troy Weights.

	gr.	dwt.	oz.	lb.
Grain	1			
Pennyweight	24	1		
Ounce	480	20	1	
Pound	5 760	240	12	1

NOTES. 1. Troy weight is chiefly employed in weighing gold, silver, and precious stones.

2. Apothecaries, in compounding medicines, divide the ounce (℥) into 8 *drachms* (ℨ) and the drachm into 3 *scruples* (℈), so that 1 scruple = 20 grains.

3. 480 *minims* = 1 *fluid-ounce*, 20 fluid-ounces = 1 pint.

APPENDIX. 173

II.

UNITED STATES WEIGHTS AND MEASURES.

THE fundamental unit of time in the United States, as in England, is the mean solar day. It is also divided, as in England, into hours, minutes, and seconds.

The United States standards of length and mass are copies of old English standards, and are very nearly the same as the present Imperial standards of Great Britain.

Careful comparison has shown that the United States actual standard yard (a brass scale made by Troughton of London, in 1813, and deposited in the Office of Weights and Measures in Washington) is equal to 1·000024 Imperial yards, and that the United States actual standard of mass (a Troy pound of brass, made by Kater in 1827, and deposited in the United States Mint) is equal to 0·99999986 Imperial Troy pounds. The same ratio 0·99999986 also exists between the United States and Imperial Avoirdupois pounds.

The United States Gallon, or standard unit of liquid measure, is the old wine gallon of England, and contains 231 cubic inches.

The United States Bushel, or standard unit of dry measure, is the British Winchester bushel (formerly a standard in England), and contains 2150·42 cubic inches.

The relative value of the Imperial and United States standards of capacity is as follows :—

Comparative Values of English and U. S. Units of Capacity.
1 Imperial Gallon (277·274 cub. in.) = 1·2001 U. S. Gallons.
1 U. S. do. (231·000 cub in.) = 0·8331 Imperial Gallons.
1 Imperial Bushel (2218·192 cub. in.) = 1·0315 U. S. Bushels.
1 U. S. do. (2150·420 cub. in.) = 0·96945 Imperial Bushels.

The ratio of the United States and Imperial gallons is very nearly as 5 : 6 ; and that of the two bushels nearly as 16 : 17.

With one exception, the United States *tables* of Weights and Measures are the same as the English tables which have been

APPENDIX.

given. This exception occurs in Avoirdupois Weight, in which the United States quarter is equal to 25 lbs., and the United States ton therefore is equal to 2000 lbs.

NOTES. 1. The English (long or gross) ton of 2240 lbs. is still used in estimating English goods in the United States Custom Houses, in selling coal at wholesale from the Pennsylvania mines, and in the wholesale iron and plaster trade.

2. *In liquid measure* 63 United States gallons = 52½ Imperial gallons = 1 hogshead.

3. 1 United States *fluid-ounce* = 1 fluid-ounce and 20 minims, Imperial measure; and 16 United States fluid-ounces = 1 United States pint.

4. *The United States barrel* contains from 28 to 32 United States gallons.

5. *In dry measure* the half-peck, sometimes called the *dry* gallon, contains 268·8 cubic inches, so that liquid and dry measures, of the same name (for example, the *quart*), stand to each other in value as 231 : 268·8.

6. The bushel, *heaped measure*, contains about 2750 cubic inches, or rather more than 5 pecks.

III.
THE METRIC SYSTEM OF WEIGHTS AND MEASURES.

The fundamental units of Time, Length, and Mass, respectively, are the mean solar day, the metre, and the kilogramme.

The **Mean Solar Day** has been already defined (see English Weights and Measures).

All civilized nations, in fact, employ the mean solar day as the fundamental unit of time, and divide it into hours, minutes, and seconds, as in England and the United States.

Its universal use is due to the fact that it combines, in an unrivalled degree, the cardinal virtues of a fundamental unit of measurement. Its magnitude is determined, not by legislators or men of science, but by nature; its duration can be easily ascertained to a high degree of precision; it has not changed since the time of Hipparchus (150 B. C.) by so much as $\frac{1}{100}$ of a second; and finally, the purposes of life and the actions of men are so largely dependent upon the position of the sun that a measure of time derived from his motion is vastly more convenient than any other.

The **Metre**, or standard of Length, is the distance between the ends of a certain platinum bar made by Borda, and preserved in the Archives de l'État in Paris, the bar being at the temperature of melting ice. The metre was intended to be an exact ten-millionth part of a quadrant of the earth's meridian, but according to the most trustworthy measurements the actual standard metre (Borda's platinum rod) is less than its intended value by an amount not exceeding $\frac{1}{5000}$ of a metre, or about $\frac{1}{180}$ of an inch.

The **Kilogramme**, or standard of Mass, is the quantity of matter in a certain platinum weight made by Borda, and deposited in the Archives de l'État in Paris. It was intended to be (and is very nearly) equal to the mass of one cubic decimetre of distilled water at the point of maximum density (about 4° C.).

The metre and the kilogramme derive their authority as standards from a law of the French Republic in 1795. Of the prototype standards, kept at Paris, numerous copies have been taken, which, after having been compared with the originals with the utmost precision of which modern science is capable, have been made standards of reference and verification in the various countries in which the Metric System has been adopted.

APPENDIX.

The Metric System of Weights and Measures is a *system* in the true sense of that word, and the simple mode in which all the other units are derived from the standards is in striking contrast with the want of system and troublesome numerical relations existing in English Weights and Measures.

Nearly all the derived units in the Metric System are obtained by the application of two very simple principles, viz. *decimal multiplication and division*, and *the use of squares and cubes of linear units as units of surface and of volume respectively*. Hence each unit of length, and also of mass, is 10 times larger than the next smaller unit in order, each unit of surface 100 times larger, and each unit of volume 1000 times larger.

Names for all the units are formed from the roots *metre* and *gramme*, by employing the Greek prefixes, *deca* (10), *hecto* (100), *kilo* (1000), for the multiples, and the Latin prefixes, *deci* (0·1), *centi* (0·01), *milli* (0·001), for the sub-multiples. In the measures of surface and of volume the words *square* and *cubic* respectively are also employed. Thus, 1 *centimetre* is equal to 0·01 of a metre, 1 *cubic decimetre* is equal to 0·001 of a cubic metre, 1 *kilogramme* is equal to 1000 grammes, &c.

Four units have received distinct names, viz. : —
The **Are**, which is equal to 1 square decametre.
The **Stere**, which is equal to 1 cubic metre.
The **Litre**, which is equal to 1 cubic decimetre.
The **Tonne**, which is equal to 1000 kilogrammes.

The table opposite exhibits the names, abbreviations, and relative values of nearly all the measures in the Metric System.

NOTES 1. The Are is employed as a unit of land measure; the Stere as a unit for measuring cord-wood; and the Litre as a unit for measuring fluids, and dry substances in small pieces, as grain, fruit, salt, &c.
2. Decimal multiples and divisions of the litre are in use, and are named by employing Greek and Latin prefixes in the same way as in measures of length; thus, *decilitre, kilolitre,* &c. Similarly, 100 ares is called a *hectare*, $\frac{1}{100}$ of an are a *centiare*, 10 steres a *decastere*, $\frac{1}{10}$ of a stere a *decistere*, 10000 metres a *myriametre*, 10000 grammes a *myriagramme*.
3. Binary derivatives, as in England, are much employed on account of their practical convenience; e. g. the *demi-litre*, the *double hectogramme*, &c., &c.
4. The cubic metre, when employed in France as a measure of the capacity of ships, is called a *tonneau*. The word *tonneau* also denotes, in the French navy, the weight of a cubic metre of water.
5. 100 kilogrammes is called a *quintal*.

TABLE OF WEIGHTS AND MEASURES IN THE METRIC SYSTEM.

MEASURES OF

Length.	Surface.	Volume.	Mass.
10	100	1000	10
Kilometre (km.)*	Square kilometre (km.^{2})	Cubic kilometre (km.^{3})	**Kilogramme** (kg.)*
Hectometre (hm.)	Square hectometre (hm.^{2})*	Cubic hectometre (hm.^{3})	Hectogramme (hg.)
Decametre (dcm.)	Square decametre (dcm.^{2})*	Cubic decametre (dcm.^{3})	Decagramme (dcg.)
Metre (m.)*	Square metre (m.^{2})*	Cubic metre (m.^{3})*	Gramme (g.)*
Decimetre (dm.)*	Square decimetre (dm.^{2})	Cubic decimetre (dm.^{3})*	Decigramme (dg.)
Centimetre (cm.)*	Square centimetre (cm.^{2})	Cubic centimetre (cm.^{3})*	Centigramme (cg.)
Millimetre (mm.)*	Square millimetre (mm.^{2})	Cubic millimetre (mm.^{3})	Milligramme (mg.)*

The number at the head of each column denotes the ratio of any unit in that column to the unit next below it. | 1 Tonne = 1000 Kilogrammes.

NOTE. — In the matter of abbreviations, as in that of pronunciation and of orthography, usage is not at present uniform. The abbreviations here given agree substantially with French usage, and are believed to be the most simple, the most appropriate, and the most convenient. The measures most in use are marked thus (*). The standards of length and of mass are printed in heavy-face type.

APPENDIX.

TABLE OF RELATIVE VALUES OF ENGLISH AND METRIC MEASURES.

English in Metric.	Value.	Metric in English.	Value.
1 English inch in centimetres	2·5399772	1 centimetre in English inches	0·39370432
1 English foot in metres	0·30479727	1 metre in English feet	3·28086933
1 English yard in metres	0·91439180	1 metre in English yards	1·09362311
1 English statute mile in kilometres	1·6093295	1 kilometre in English statute miles	0·62137677
1 square foot in centiares	0·09290137	1 centiare in square feet	10·764104
1 acre in hectares	0·40467838	1 hectare in acres	2·4710982
1 square mile in hectares	258·99416	1 hectare in square miles	0·00386109
1 cubic foot in cubic metres (steres)	0·02831608	1 cubic metre in cubic feet	35·315617
1 cord in steres	3·6244589	1 stere in cords	0·2759033
1 Imperial gallon in litres	4·5435845	1 litre in Imperial gallons	0·2200906
1 U. S. gallon in litres	3·785	1 litre in U. S. gallons	0·264
1 U. S. bushel in litres	35·24	1 litre in U. S. bushels	0·02838
1 Imperial quart in litres	1·1358961	1 litre in Imperial quarts	0·8803624
1 U. S. liquid quart in litres	0·94625	1 litre in U. S. liquid quarts	1·056
1 U. S. dry quart in litres	1·10125	1 litre in U. S. dry quarts	0·90816
1 grain in grammes	0·06479895	1 gramme in grains	15·43234874
1 pound (Avoir.) in kilogrammes	0·45359265	1 kilogramme in pounds (Avoir.)	2·20462125
1 pound (Troy) in kilogrammes	0·37324195	1 kilogramme in pounds (Troy)	2·67922721
1 foot-pound (Avoir.) in kilogramme-metres	0·1382538	1 kilogramme-metre in foot-pounds (Avoir.)	7·2330743

APPENDIX. 179

IV.

MATHEMATICAL DATA AND FORMULAE.

ALGEBRA.

Arithmetical Progression.

(1) Let a denote the first term, l the last term, d the common difference, n the number of terms, s the sum of the terms; then, —

$$l = a + (n-1)d; \qquad s = n\,\frac{a+l}{2}.$$

Geometrical Progression.

(2) Let the same notation as above be used, except that in this case r denotes the common ratio; then, —

$$l = a r^{n-1}; \qquad s = \frac{rl-a}{r-1}.$$

GEOMETRY.

(3) π (ratio of circumf'nce of circle to diameter) $= 3{\cdot}14159$
(4) Circumference of a circle (radius r) $= 2\pi r$
(5) Area of a circle (radius r) $= \pi r^2$
(6) Surface of a sphere (radius r) $= 4\pi r^2$
(7) Surface of ellipse (semi-axes a and b) $= \pi a b$
(8) Surface of right cylinder (height h, base πr^2) $= 2\pi r h$
(9) Surface of right cone (height h, base πr^2) $= \pi r \sqrt{r^2 + h^2}$
(10) Volume of a sphere (radius r) $= \tfrac{4}{3}\pi r^3$
(11) Volume of ellipse (semi-axes a and b) $= \tfrac{4}{3}\pi a b c$
(12) Volume of right cylinder (height h, base πr^2) $= \pi r^2 h$
(13) Volume of right cone (height h, base πr^2) $= \tfrac{1}{3}\pi r^2 h$
(14) Sum of the three angles of a triangle $= 180°$
(15) Area of a triangle of altitude h and base $b = \tfrac{1}{2} b h$.
(16) Two triangles, which have equal bases and equal altitudes, are equal.

APPENDIX.

(17) Two triangles are similar ;
   (a) If they are equiangular with respect to each other.
   (b) If they have their homologous sides proportional.
   (c) If they have an angle of one equal to an angle of the other, and the sides including these angles proportional.
(18) The perpendicular upon the hypothenuse of a right triangle from the vertex of the right angle, divides the triangle into two triangles which are similar to each other and to the whole triangle.
(19) If a perpendicular be erected from any point on the diameter of a circle meeting the circumference at a point A, and chords be drawn from A to the extremities of the diameter ; then, —
   (a) The perpendicular is a mean proportional between the segments of the diameter.
   (b) Either chord is a mean proportional between the diameter and the adjacent segment.
(20) The square described upon the hypothenuse of a right triangle is equivalent to the sum of the squares described on the other two sides.
(21) Similar triangles (or polygons) are to each other as the squares of their homologous sides.
(22) The sum of the plane angles which form a solid angle is always less than four right angles.
(23) The area of the surface described by a line revolving about another line on the same plane as an axis is the product of the revolving line by the circumference described by its middle point.

PLANE TRIGONOMETRY.

Definitions of the Functions of an Angle.

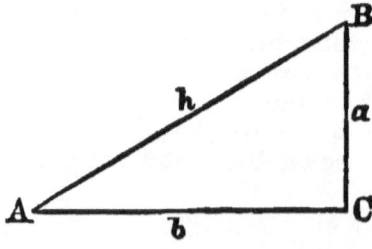

Let h denote the hypothenuse of a right triangle, a the perpendicular, b the base, A and B the angles opposite a and b respectively ; then, —

APPENDIX. 181

(24) $\begin{cases} \sin A = \dfrac{a}{h}, & \tan A = \dfrac{a}{b}, & \sec A = \dfrac{h}{b}, \\ \cos A = \dfrac{b}{h}, & \cot A = \dfrac{b}{a}, & \operatorname{cosec} A = \dfrac{h}{a}. \end{cases}$

(25) *Values of the Functions of particular Angles.*

Angle.	Arc.	sin	cos	tan	cot	sec	cosec
0°	0	0	1	0	∞	1	∞
30°	$\tfrac{1}{6}\pi$	$\tfrac{1}{2}$	$\tfrac{1}{2}\sqrt{3}$	$\sqrt{\tfrac{1}{3}}$	$\sqrt{3}$	$2\sqrt{\tfrac{1}{3}}$	2
45°	$\tfrac{1}{4}\pi$	$\sqrt{\tfrac{1}{2}}$	$\sqrt{\tfrac{1}{2}}$	1	1	$\sqrt{2}$	$\sqrt{2}$
60°	$\tfrac{1}{3}\pi$	$\tfrac{1}{2}\sqrt{3}$	$\tfrac{1}{2}$	$\sqrt{3}$	$\sqrt{\tfrac{1}{3}}$	2	$2\sqrt{\tfrac{1}{3}}$
90°	$\tfrac{1}{2}\pi$	1	0	∞	0	∞	1
120°	$\tfrac{2}{3}\pi$	$\tfrac{1}{2}\sqrt{3}$	$-\tfrac{1}{2}$	$-\sqrt{3}$	$-\sqrt{\tfrac{1}{3}}$	-2	$2\sqrt{\tfrac{1}{3}}$
135°	$\tfrac{3}{4}\pi$	$\sqrt{\tfrac{1}{2}}$	$-\sqrt{\tfrac{1}{2}}$	-1	-1	$-\sqrt{2}$	$\sqrt{2}$
150°	$\tfrac{5}{6}\pi$	$\tfrac{1}{2}$	$-\tfrac{1}{2}\sqrt{3}$	$-\sqrt{\tfrac{1}{3}}$	$-\sqrt{3}$	$-2\sqrt{\tfrac{1}{3}}$	2
180°	π	0	-1	0	∞	-1	∞

Useful Formulae, in which α and β denote any two Angles.

(26) $\qquad \sin^2 \alpha + \cos^2 \alpha = 1.$

(27) $\qquad \tan \alpha = \dfrac{\sin \alpha}{\cos \alpha} = \dfrac{1}{\cot \alpha}.$

(28) $\qquad \sec^2 \alpha = 1 + \tan^2 \alpha.$

(29) $\qquad \operatorname{cosec}^2 \alpha = 1 + \cot^2 \alpha.$

(30) $\qquad \text{versed sin } \alpha = 1 - \cos \alpha.$

APPENDIX.

(31) $\quad \sin 2a = 2 \sin a \cos a.$
(32) $\quad \cos 2a = \cos^2 a - \sin^2 a.$

(33) $\quad \tan 2a = \dfrac{2 \tan a}{1 - \tan^2 a}.$

(34) $\quad \sin (a \pm \beta) = \sin a \cos \beta \pm \cos a \sin \beta.$
(35) $\quad \cos (a \pm \beta) = \cos a \cos \beta \mp \sin a \sin \beta.$

(36) $\quad \tan (a \pm \beta) = \dfrac{\tan a \pm \tan \beta}{1 \mp \tan a \tan \beta}.$

(37) $\quad \cot (a \pm \beta) = \dfrac{\cot \beta \cot a \mp 1}{\cot \beta \pm \cot a}.$

(38) $\quad \sin a + \sin \beta = 2 \sin \tfrac{1}{2}(a+\beta) \cos \tfrac{1}{2}(a-\beta).$
(39) $\quad \sin a - \sin \beta = 2 \cos \tfrac{1}{2}(a+\beta) \sin \tfrac{1}{2}(a-\beta).$
(40) $\quad \cos a + \cos \beta = 2 \cos \tfrac{1}{2}(a+\beta) \cos \tfrac{1}{2}(a-\beta).$
(41) $\quad \cos a - \cos \beta = -2 \sin \tfrac{1}{2}(a+\beta) \sin \tfrac{1}{2}(a-\beta).$

In any plane triangle, let a, b, c, denote the sides, A, B, C, the angles respectively opposite the sides, and let $s = \tfrac{1}{2}(a+b+c)$; then, —

(42) $\quad \dfrac{a}{\sin A} = \dfrac{b}{\sin B} = \dfrac{c}{\sin C}.$

(43) $\quad \dfrac{a+b}{a-b} = \dfrac{\tan \tfrac{1}{2}(A+B)}{\tan \tfrac{1}{2}(A-B)}.$

(44) $\quad \begin{cases} a^2 = b^2 + c^2 - 2bc \cos A, \\ b^2 = c^2 + a^2 - 2ca \cos B, \\ c^2 = a^2 + b^2 - 2ab \cos C. \end{cases}$

(45) $\quad \sin \tfrac{1}{2} a = \sqrt{\dfrac{(s-b)(s-c)}{bc}}.$

(46) $\quad \cos \tfrac{1}{2} a = \sqrt{\dfrac{s(s-a)}{bc}}.$

(47) $\quad \tan \tfrac{1}{2} a = \sqrt{\dfrac{(s-b)(s-c)}{s(s-a)}}.$

APPENDIX.

ANALYTIC GEOMETRY.

General Equation of the Straight Line.

(48) $$A x + B y + C = 0.$$

In the following equations of the straight line, a and b denote the intercepts of the line on the axes of x and y respectively, m the tangent of the angle which the line makes with the axis of x, p the perpendicular on the line from the origin, a the angle which this perpendicular makes with the axis of x.

(49) $$\frac{x}{a} + \frac{y}{b} = 1.$$

(50) $$y = m x + b.$$
(51) $$x \cos a + y \sin a = p.$$

(52) Two straight lines, the equations of which are $y = m x + b$, and $y = m' x + b'$, are *parallel* when $m = m'$, and are *perpendicular* when $m\, m' + 1 = 0$.

General Equation of a Conic Section.

(53) $$a x^2 + 2 h x y + b y^2 + 2 g x + 2 f y + c = 0.$$
If $h^2 - a b < 0$, locus is an *ellipse*,
If $h^2 - a b < 0$, and $a = b$, locus is a *circle*,
If $h^2 - a b > 0$, locus is an *hyperbola*,
If $h^2 - a b = 0$, locus is a *parabola*.

In the following equations of the conic sections, r denotes the radius of the circle, a and b the semi-axes of the ellipse or hyperbola, $e = \sqrt{\dfrac{a^2 - b^2}{a^2}}$, and is the measure of the eccentricity of the ellipse or hyperbola, $p = $ the parameter of the parabola (that is, the double ordinate at the focus), m the distance from the vertex of a parabola to the focus (or *focal distance*).

Circle.

(54) Centre at point $(a\ \beta)$; $(x - a)^2 + (y - \beta)^2 = r^2$.
(55) Origin at centre; $x^2 + y^2 = r^2$.
(56) Axis of x diameter, origin on circumf'nce; $x^2 + y^2 = 2 r x$.
(57) Tangent at point $(x'\, y')$, origin at centre; $x x' + y y' = r^2$.

APPENDIX.

Ellipse.

(58) Origin at centre . . . . $\dfrac{x^2}{a^2}+\dfrac{y^2}{b^2}=1.$

(59) Origin at vertex . . . . $y^2=\dfrac{b^2}{a^2}(2\,a\,x-x^2).$

(60) Pole at centre . . . . $\rho^2=\dfrac{b^2}{1-e^2\cos^2\theta}.$

(61) Pole at focus . . . . $\rho=\dfrac{b^2}{a\,(1+e\cos\theta)}.$

(62) Tangent at point $x'\,y'$ . . . . $\dfrac{x\,x'}{a^2}+\dfrac{y\,y'}{b^2}=1.$

(63) Focal radii make equal angles with the tangent.

Hyperbola.

The equations of the hyperbola are the same as those of the ellipse, except that b^2 is negative.

Parabola.

(64) Origin at vertex . . . . $y^2=p\,x=4\,m\,x.$

(65) Pole at focus . . . . $\rho=\dfrac{2\,m}{1-\cos\theta}.$

(66) Tangent at point $x'\,y'$ . . . . $2\,y\,y'=p\,(x+x').$
(67) The subtangent is bisected at the vertex.
(68) The subnormal is constant and equal to $\tfrac{1}{2}\,p.$
(69) The point where any tangent cuts the axis of x, and its point of contact, are equally distant from the focus.
(70) Any tangent makes equal angles with the axis of x and the focal radius vector.

Analytic Geometry of Three Dimensions.

If l, m, n, denote the cosines of the three angles which a straight line makes with three rectangular axes, or *direction cosines*; then, —
(71) $\qquad\qquad l^2+m^2+n^2=1.$

If θ denote the mutual inclination of two lines $(l\ m\ n)$ $(l'\ m'\ n')$; then, —
(72) $\qquad\qquad \cos\theta=l\,l'+m\,m'+n\,n'.$

V.
PHYSICAL DATA AND TABLES.
Table I.
Value of the Acceleration of Gravity.

Place.	Latitude.	Value of g in Metres.	Seconds Pendulum in Metres.
Spitzbergen............	79° 49' 58" N.	9·83141	0·99613
Stockholm............	59 20 34 "	9·81946	0·99492
Königsberg...........	54 42 12 "	9·81443	0·99441
London	51 30 48 "	9·81111	0·99409
Paris	48 50 14 "	9·80979	0·99394
Isle Rawak	0 1 34 S.	9·78206	0·99113
Isle de France	20 9 23 "	9·78917	0·99185
Cape of Good Hope	33 55 15 "	9·79696	0·99264
Cape Horn............	55 51 20 "	9·81650	0·99462
New Shetland	62 56 11 "	9·82253	0·99523

The value of g is usually determined experimentally by pendulum observations. It may be calculated with sufficient accuracy by the following formula, —

$$g = G\,(1 - 0{\cdot}0025659 \cos 2\lambda)\left(1 - 1{\cdot}32\,\frac{h}{r}\right),$$

in which G = value of g for the latitude $45°$ = 32·1703 feet = 9·80533 metres, r = radius of the earth = 20 886 852 feet = 6 366 198 metres, and h = height of the place in feet or metres above the level of the sea.

Table II.
Dimensions of Dynamical Units,
T denoting a time, L a length, and M a mass.

	Velocity.	Acceleration.	Momentum.	Force.	Energy.
Dimension.	$\dfrac{L}{T}$	$\dfrac{L}{T^2}$	$\dfrac{ML}{T}$	$\dfrac{ML}{T^2}$	$\dfrac{ML^2}{T^2}$

$mass \times velocity = mass \times acceleration \times time = force \times time.$

TABLE III.

Specific Gravities of Solids and Liquids, referred to that of Distilled Water at 4° C. as a standard.

Metals.		Rocks.	
Platinum, hammered	22·060	Granite	2·65 to 2·75
Gold, "	19·350	Sandstone	2·25 to 2·65
Silver, "	10·510	Limestone	2·60 to 2·70
Copper "	8·900	Marble	2·65 to 2·75
Lead, cast	11·350	Slate	2·84
Tin, cast	7·290	Porcelain Clay	2·21
Zinc	7·190	Sand, dry	1·42
Iron, cast	7·250		
Iron, wrought	7·780	*Various Substances.*	
Steel, soft	7·830	Diamond	3·530
Steel, tempered	7·810	Glass, flint	3·000
Brass, cast	7·820	Glass, crown	2·520
Brass, sheet	8·390	Sulphur	2·086
Bronze, statuary	8·950	Brick	2·000
Bronze, gun metal	8·460	Coal, anthracite	1·800
Aluminum	2·600	Coal, bituminous	1·270
		Wax	0·960
Woods.		Ice	0·918
Lignum Vitae	1·333		
Box, Dutch	1·328	*Liquids.*	
Box, French	0·912	Mercury	13·596
Ebony, American	1·331	Bromine	2·966
Oak, just felled	1·113	Sulphuric Acid, com'cial	1·841
Oak, seasoned	0·743	Nitric Acid, "	1·500
Mahogany, Spanish	1·063	Muriatic Acid "	1·200
Beech	0·852	Chloroform	1·492
Ash	0·845	Milk	1·031
Maple	0·790	Sea-water	1·025
Elm	0·673	Proof Spirit	0·920
Chestnut	0·565	Absolute Alcohol	0·795
Red Pine	0·657	Sulphuric Ether	0·715
White Pine	0·551	Olive Oil	0·915
Larch	0·530	Oil of Turpentine	0·870
Cork	0·240	Naphtha	0·840

APPENDIX. 187

TABLE IV.

Specific Gravities of Gases and Vapors, referred to that of Dry Air at the same Temperature and Pressure as the Standard.

Oxygen	1·106	Vapor of Water	0·623
Hydrogen	0·069	Carbonic Oxide	0·967
Nitrogen	0·972	Carbonic Acid	1·525
Chlorine	2·470	Ether	2·570
Ammonia	0·590	Marsh Gas	0·555
Nitrous Oxide	1·520	Coal Gas	0·420 to 0·520

Weight of 1 litre of dry air at 0° C. and 76 centimetres pressure at Paris = 1·293 grammes. Weight of 1 cubic foot of dry air at 32° F. and 29·905 inches pressure at London = 565 grains, or about 1·25 ounces avoirdupois.

BAROMETRIC FORMULAE FOR FINDING DIFFERENCES OF LEVEL.

In these formulae X denotes difference of level, h height of barometer at lower station reduced to 0° C., h' the same for higher station, t temperature at lower station, t' the same for higher station (t and t' being expressed in Fahrenheit degrees in I. and in Centigrade degrees in II.), λ the latitude of the stations. The atmosphere is supposed to be in a mean hygrometric condition.

The second and simpler value of X, given in both I. and II., will suffice for heights not exceeding 3300 feet, or 1000 metres.

I. X (in feet) $= 60384 \log \dfrac{h}{h'} \left[1 + \dfrac{t + t' - 64}{900} \right] \times$
$(1 + 0·00256 \cos 2\lambda)$
$= 52500 \dfrac{h - h'}{h + h'} \left[1 + \dfrac{t + t' - 64}{900} \right].$

II. X (in metres) $= 18400 \log \dfrac{h}{h'} \left[1 + \dfrac{2(t + t')}{1000} \right] \times$
$(1 + 0·00256 \cos 2\lambda)$
$= 16000 \dfrac{h - h'}{h + h'} \left[1 + \dfrac{2(t + t')}{1000} \right].$

APPENDIX.

THE ASCENSIONAL FORCE OF A BALLOON.

Let V denote space occupied by the gas, v volume of the solid parts of the balloon, w weight of the same, w' weight of the aeronauts, a and a' weights of unit volumes of air and the gas respectively at 0° C. and 76 centimetres pressure, h air-pressure at time of ascension, h' air-pressure at the altitude where the balloon is in equilibrium, P ascensional force; then, changes of temperature being neglected, —

I. $$P = (V + v)\frac{a\,h}{76} - V\frac{a'\,h}{76} - w - w'.$$

II. $$(V + v)\frac{a\,h'}{76} = V\frac{a'\,h}{76} + w + w'.$$

This last formula, taken in connection with the barometric formulae before given, enables us to find roughly the size of the balloon necessary for reaching a given height with a given load. It is, however, of but small practical value. In the upper regions of the air, the pressure of the gas in the balloon may become so much greater than that of the external air, that the aeronaut will find it prudent to allow some of the gas to escape. Moreover, aeronauts are often obliged to throw out ballast under circumstances which cannot be foreseen. The changes of temperature, also, in different air-strata, which are very considerable, are not taken into account in the formula, and cannot be allowed for, being largely determined by the winds and atmospheric currents which happen to prevail at the time.

THE END.

www.ingramcontent.com/pod-product-compliance
Lightning Source LLC
Chambersburg PA
CBHW030820190426
43197CB00036B/621